STARS

A Month-by-Month Tour of the Constellations

MIKE LYNCH

PUBLICATIONS
Adventure
an imprint of Adventure**KEEN**

DEDICATION

To my wife Kathy, my daughter Angie, my son Shaun, my mother Eileen; Tanja Michaels, who's given me wonderful assistance for years; and a dear friend Jennifer Evans, who's a true up-and-coming star. I'd also like to thank my Brother in Christ, Dean Koenig, the owner of Starizona in Tucson, Arizona, for all his tremendous help and advice over the years. A very special dedication to my father Donald Lynch, my sister Mary Lynch Veldey, and my brother Joe Lynch, who have made the ultimate trip to the heavens.

Constellation artwork by Angela Lynch

Front and back cover photos by Pozdeyev Vitaly/ Shutterstock.com

Photo Credits: Moon photos on pages 34 and 42 courtesy of Jennifer Evans, author photo by Kathy Lynch, all other photos by Mike Lynch

Edited by Brett Ortler and Andrew Mollenkof

Book and cover design by Jonathan Norberg

10 9 8 7 6 5 4 3 2 1

Stars: A Month-by-Month Tour of the Constellations
Copyright © 2012 and 2024 by Mike Lynch
All rights reserved
Printed in China
Published by Adventure Publications
Distributed by Publishers Group West
LCCN 2023050182 (print); 2023050183 (ebook)
ISBN 978-1-64755-419-4 (pbk.); 978-1-64755-420-0 (ebook)

Adventure Publications
An imprint of AdventureKEEN
310 Garfield Street South
Cambridge, MN 55008
(800) 678-7006
adventurepublications.net

JELLYFISH NEBULA (right)

TABLE OF CONTENTS

READY TO MAKE THE STARS YOUR OLD FRIENDS?

I've been old friends with the night sky for over 50 years, and now you can be too. I created this book of star maps and constellation charts to give you a guided tour of the night sky. If you're brand-new to stargazing, I hope this book helps you get acquainted with the cosmos. If you're already an avid stargazer, I hope this offering gets you even more excited. Whatever your level of interest, I sure hope you enjoy astronomy and stargazing as much as I have.

ASTRONOMY 101

What follows will certainly not be a comprehensive course in astronomy. For that, there are many great books, films, and classes where you can learn a heck of a lot more. With that said, let me try to give you the bare essentials.

WHAT ARE STARS?

Stars are basically big balls of gas that are primarily made up of hydrogen and some helium, but they include many other elements as well. Stars are born in large clusters and form from loose clouds of hydrogen gas. These clouds are laced with heavier elements that were created in very massive stars that exploded eons ago in what astronomers call supernovas.

These gas clouds formed dense clumps and eventually combined into much larger balls of gas thanks to gravity. When they became massive enough, gravitational pressure caused the core temperatures of these "protostars" to reach millions of degrees. This set off a nuclear chain reaction inside the stellar cores, and individual hydrogen atoms began colliding with one another with enough force that they fused into helium atoms, which are heavier. There are zillions of such collisions every second. Some of the hydrogen in them is converted into light and radiation, which makes its way out of the gargantuan ball of gas—now a full-fledged star—to help light up the night.

For stars like our sun, normal nuclear fusion can keep going for well over 10 billion years. Really massive stars are gas guzzlers and go through their nuclear fuel much faster, some in as little as a few million years. The details are a little complicated, but when a star begins to run out of hydrogen in its core, the star begins to expand, eventually turning into a huge red giant that lasts for a few more billion years.

After that, less-massive stars (like our sun) turn into small white dwarf stars and eventually flicker out. Much larger stars have a different fate, becoming unstable and exploding violently in a supernova. The extreme conditions in supernovas "cook up" heavy elements like gold, silver, and uranium and spew them out in all directions, where they become components of new stars and planets. You even have bits of these exploded stars in your own body.

What's left of these exploded stars continues to collapse, forming hyper-dense pulsars and neutron stars, objects that are so dense that one tablespoon would weigh hundreds of millions of tons on Earth! Stars that are massive enough may collapse into black holes, which have so much gravitational pull that not even light can escape.

DISTANCES TO THE STARS

Talking about stellar distances in miles can get really cumbersome. Instead, stellar distances are measured in light-years. Light travels at just over 186,000 miles per second. How fast is that? If you had a jet airliner that could make the round trip from Los Angeles to New York and back 33 times in one second, you'd be traveling at the speed of light. Over the course of a year, light travels just under 6 trillion miles; this distance is called a light-year. The stars that you see in the night sky can range in vast distances from you; some are under 10 light-years from Earth, while others are thousands of light-years away. Also, keep in mind that when you look into the night sky, you're actually looking back in time: after all, even light needed a lot of time to travel those incredible distances. For example, if you gaze at a star 100 light-years away, you're seeing that star as it was 100 years ago, as the light it is emitting now is just beginning its journey. If a star is 1,000 light-years away, you're seeing it as it was a millennium ago.

GALAXIES

All the stars you see in the sky are part of our home Milky Way Galaxy, which has at least 200 to 300 billion stars. Some astronomers now believe, though, that planets may easily exceed the number of stars in the Milky Way. There are several types of galaxies in the known universe. The Milky Way is considered a spiral galaxy because it has "arms" that radiate out like a pinwheel. The Milky Way has a diameter of over 100,000 light-years but is only about 1,000 light-years thick, although the central bulge is about 12 times thicker. Most of the stars in the galaxy are located in that bulge (the galactic center), which is also home to a massive black hole estimated to be over 4 million times more massive than our sun. The rest of the Milky Way's stars—including our sun—are located in one of the spiral arms, relatively far away from the core. Sometimes, especially in the summer, you can see a narrow ribbon of milky white light that stretches across the celestial dome. When you see that, you're looking into the galaxy's thinner plane. There are so many stars there (and they are so far away) that all you see is their combined glow.

Here's something to ponder: Most of the mass of our galaxy is invisible and no one really knows what it is. For now, we call it dark matter, and it is one of the biggest mysteries in astronomy. And if that didn't blow your mind, just consider that our Milky Way is only one of billions and billions of galaxies in the known universe, and the known universe has a diameter of over 40 billion light-years. Even weirder, the universe is expanding, and its galaxies are generally moving away from each other, and their movement is accelerating. There are so many things to consider as you make the stars your old friends!

CONSTELLATIONS

When most people think of constellations, they often think of "pictures" in the heavens that represent people, animals, and objects. For many ancient cultures, constellations served a variety of purposes: they were rough visual aids, helped mariners navigate, and enabled cultures to depict their stories and mythology. Unfortunately, most constellations don't look at all like what they're supposed to represent; to see the resemblances, you often need to use your imagination.

More recently, astronomers have divided the entire sky into 88 constellations. These constellations all have parallel and perpendicular boundaries, creating a rough map of the sky. All of the stars in the sky belong to one constellation or another. Constellations come in a variety of sizes and can be bright or dim. At any given time, there are about 44 constellations in the sky, and through the course of the year, you can see about 65–70 constellations, depending on where you live.

In this book, I feature two or three constellations that are prominent in a given month. For each, I've

SOUL NEBULA

included drawings of what these constellations allegedly represent. I've also featured some other objects that you can target with most small- to moderate-size telescopes, including star clusters, nebulae, and galaxies. I also discuss some of the Greek and Roman mythology commonly associated with each constellation.

USING THE MONTHLY STAR MAPS

This book features 12 star maps that clearly and simply depict the night sky in each month. Ideal for use in early evening, these maps can also be used at other times of night, or in the early morning at other times of the year. (Other dates/times when the charts are representative are listed at the bottom of each map.) Since these are created for use in areas with at least some light pollution, not all stars and constellations appear on the maps. Dimmer stars and constellations have been omitted.

All seasons of the year are great for stargazing, you just need to make sure you're dressed appropriately for the season. The best way to use these maps is to get a comfortable lawn chair and sit outside, facing one of the cardinal directions. When you're ready to go, the idea is to hold the map over your head so the direction you're facing matches the compass point on the map. For example, start out facing north, and hold the star map over your head so north on the map points away from you. If you're doing it right, the "S" for south on the map will be behind you. It's best to take one direction at a time. When you're done exploring constellations in the northern sky, move your chair and face west, holding the map over your head so west on the map points away from you and toward the western horizon. Then, repeat this process for the southern and eastern sky.

When you're stargazing, be sure to give yourself a good 20 minutes or so to get used to the darkness before you start using the maps. Not only will that help you get your night vision, but it'll also help you relax and clear your mind, getting you in the right mood to get to know your universe a little more. Speaking of night vision, it's handy to have a headband flashlight with you, but make sure the light coming out of it is red. Red light doesn't cause your night vision to reset like white light does. You can buy red-tinted headband flashlights, but if you want to save money, just buy a regular one and put a piece of red electrical tape over it. With a red headband flashlight, your hands are free to hold up the star map and you can also see the maps without wrecking your night vision.

A NOTE ABOUT PLANETS, METEOR SHOWERS, AND SATELLITES

Many skywatchers are interested in spotting planets, meteor showers, satellites, and the like. Such objects are constantly on the move and it's not possible to plot them on these maps and charts. Thankfully, there are many ways to stay in the know, including subscribing to magazines like *Sky and Telescope* or *Astronomy*, or checking out their websites. (You can also learn a lot from both publications.) There are also some really cool space-related websites, such as heavens-above. com, space.com, spaceweather.com, and many others. You can also use apps like Sky Guide. I also have a web and app offering for you at lynchandthestars.com.

Now Get Out There and Truly Make the Stars Your Old Friends!

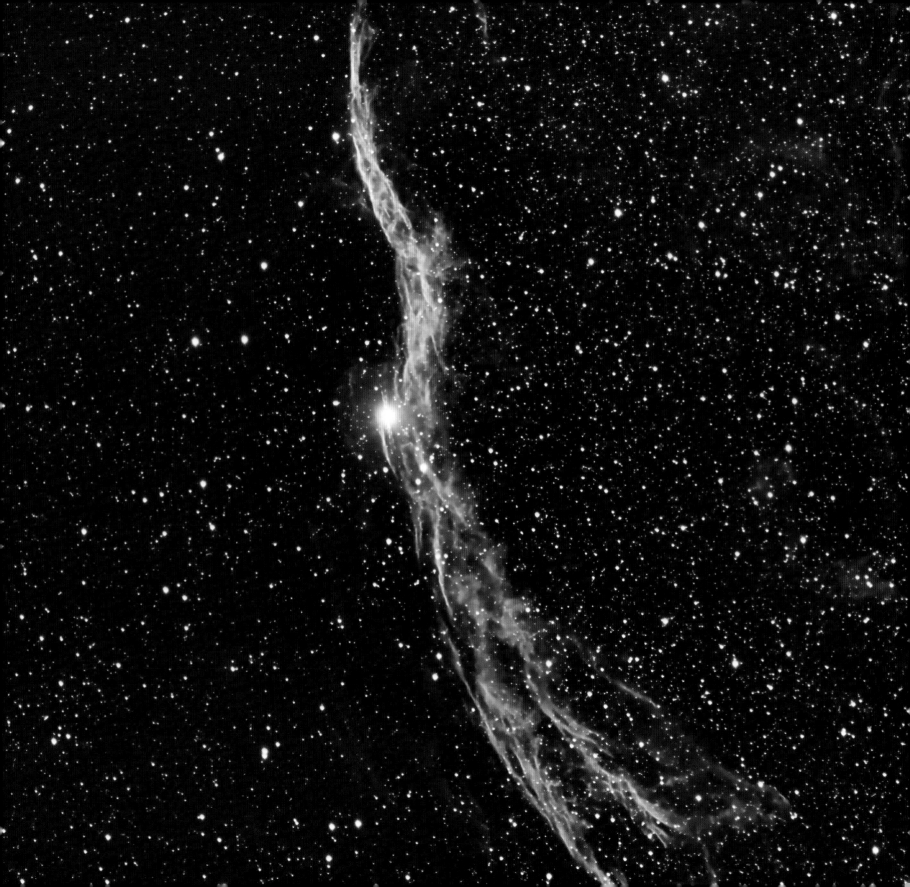

JANUARY Sky Chart

N

NE

NW

DRACO
The Dragon

Eltanin

Vega

Big Dipper
Part of Ursa Major

Mizar   Alcor

Thuban

Albireo

URSA MINOR
The Little Bear

Merak   Dubhe

Little Dipper

Alderamin

CYGNUS
The Swan

URSA MAJOR
The Big Bear

Polaris/North Star

CEPHEUS
The King

Northern Cross

LEO
The Lion

Regulus

Perseus Double
Cluster

Schedar

CASSIOPEIA
The Queen

Andromeda
Galaxy

Beehive
Cluster

AURIGA
The Chariot Driver

ZENITH

Mirfak

GEMINI
The Twins

Castor

Pollux

The
"Kids"

Capella

Algol

PERSEUS
The Hero

ANDROMEDA
The Princess

Alpheratz

Great
Square of
Pegasus

Markab

E

W

CANIS MINOR
The Little Dog

Procyon

TAURUS
The Bull

Hamal

ARIES
The Ram

PEGASUS
The Winged Horse

Betelgeuse

Pleiades
Seven Little Sisters

ORION
The Hunter

Aldebaran

Menkar

Winter
Triangle

Sirius
*The brightest
star in the sky!*

Orion's Belt

Great Orion Nebula

Mira

Rigel

Arneb

CANIS MAJOR
The Big Dog

LEPUS
The Rabbit

Nihal

SE

MS

CONSTELLATION NAME
English Name

Celestial Object

ORION and LEPUS featured on page 11

S

Sky chart is representative of 8 p.m. in January, 12 a.m. in November, and 5 a.m. in September.

The jewels of January are taking over in the cold, clear winter skies. Dress warmly and be prepared to be wowed by majestic constellations, like Orion, the Hunter, and his posse of bright stars!

These are the best times and also the worst times for stargazing. Just bundle up and think warm, and you'll be rewarded with what I think is the greatest celestial display of the year. Even if you're stargazing from the city, don't be afraid to sit back on lawn chairs and stare up into the sky, even if the neighbors think you're a little nuts . . . In fact, ask them to join you for the best show in the universe!

Believe it or not, you still have a summer constellation holding out in the northwestern sky. At the end of evening twilight, Cygnus, the Swan, is just above the horizon. Within the Swan is the famous Northern Cross, which is standing upright. Look for it as soon as darkness falls because by 8 p.m., Albireo, the star that marks the foot of the cross, will have already slipped out of view. In the western sky, autumn constellations are also visible. Look for Pegasus, the Winged Horse, with Andromeda, the Princess, in tow.

Next to the constellation Andromeda is the Andromeda Galaxy, the next-door neighbor to the Milky Way. Even though it's over 2 million light-years away, it's possible to see it with the naked eye if your sky is dark enough. Look for a tiny ghostly patch of white. This sounds crazy, but if you're having a hard time seeing it, look slightly away from it. You may see it a whole lot better if you're not looking directly at it. This is a technique amateur astronomers call "averted vision." Now if you're forced to view the Andromeda Galaxy in light-polluted skies, you should be able to see it with a small telescope

or even a good pair of binoculars. When you're looking at the Andromeda Galaxy, keep in mind that just one light-year is nearly 6 trillion miles, and this island of stars is over 2 million light-years away . . . wrap your mind around that if you can!

On the rise in the eastern sky this month is what I call "Orion and His Gang." At the center is the wonderful and distinct constellation Orion, the Hunter. There's also his surrounding gang of constellations; the group includes Taurus, the Bull; Auriga, the Chariot Driver–turned–Goat Farmer; Gemini, the Twins; and Lepus, the Rabbit, Orion's prey and nemesis.

At first glance, Orion reminds a lot of people of an hourglass or a crooked bowtie standing proudly above the eastern horizon. That outlines the trim but muscular torso of the hunter. In the middle of his massive frame are three stars in a nearly perfect row that will likely draw your attention. These luminaries outline Orion's belt. From the lower left to the upper right the stars are Alnitak, Alnilam, and Mintaka. The amazing thing about these stars is that even though they're lined up in a perfect row, they physically have nothing to do with each other. They're hundreds of light-years apart from each other. They just happen to appear in a line from our vantage point on Earth.

Orion's brightest star is Rigel, which marks the hunter's knee, and on the upper left-hand corner of the bowtie is Betelgeuse, a huge red supergiant star that has a noticeable reddish tinge to it even with the naked eye. Betelgeuse is an Arabic name that roughly translates to English as "armpit of the great one." So, when you gaze at Betelgeuse you get the pleasure of staring into one of his armpits. Oh joy!

I can pretty much guarantee that Betelgeuse is the single biggest thing you've ever seen with an unaided eye. At times, it swells to a diameter of nearly a billion miles! Our sun is not even a million miles across.

All stars are basically huge balls of hydrogen gas that shine because of nuclear fusion. During the fusion process, hydrogen deep in stellar cores is converted to helium, and a tremendous amount of light and other energy is released. Essentially, the hydrogen in a star's core is its fuel. Smaller stars, like our sun, "burn" through their hydrogen supply much more slowly than larger stars. Much larger stars, like Betelgeuse, are real gas guzzlers that go through their hydrogen much, much more rapidly. In fact, it's estimated that Betelgeuse is only 10 million years old, but already dying. Sometime in the next thousand, hundred thousand, or million years (at most), it'll blow to bits in a tremendous supernova explosion. Astronomers aren't really sure when it'll happen, but when it does, it could be as bright as the full moon for several weeks.

Supernova explosions are very important because heavier elements like iron, gold, and silver are formed in the process. They are spewed out into the galaxy to become the building blocks of future stars, planets, and maybe life.

It's fair to say that the iron in your car (what little there is), the iron in your blood, the calcium in your bones, and the gold and silver you may be wearing were all once part of a former star. You have star stuff in you! Keep that in mind when basking under the glow of Orion.

By the way, what's left of Betelgeuse or any other gigantic star after it explodes will gravitationally shrink down and become a black hole. Black holes get their moniker because they have so much gravitational force that not even light can escape. Anything that falls into a black hole is never seen or heard from again as it mashes into what physicists call a singularity.

After you check out the red giant Betelgeuse, consider Bellatrix, the bright star across from Betelgeuse. It marks the left shoulder of Orion. You won't really see much color with the naked eye, but if you point your telescope or binoculars at it, you should see that it has a very deep shade of blue to it. It's almost purple. There's really no other bright star in the sky that matches it.

Just by observing the color of a star, you can deduce quite a bit of information about it, especially its temperature. Just like flames in a campfire, stars with red or yellow flames are cooler than those with bluish flames. Bluish, purple-tinged stars like Bellatrix are much hotter than reddish-tinged stars like Betelgeuse.

Below the stars of Orion's belt, there are three more stars in a row that allegedly depict Orion's sheath. The middle "star" in the sheath is actually the Great Orion Nebula, where, even now, new stars are being born. I'll have more on the great stellar nursery in the heavens when I talk about the February night sky.

Bundle up and enjoy the January night sky, and if you can, get out into the countryside, where you can really take in the wonderful show without that nasty light pollution. Stay warm!

HORSESHEAD NEBULA

M38 CLUSTER

ORION NEBULA (right)

JANUARY Featured Constellations

Numbers under star names represent light-years

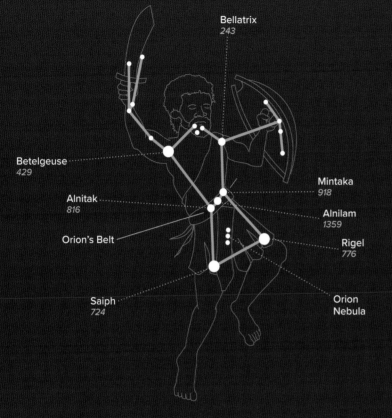

Bellatrix
243

Betelgeuse
429

Alnitak
816

Orion's Belt

Saiph
724

Mintaka
918

Alnilam
1359

Rigel
776

Orion
Nebula

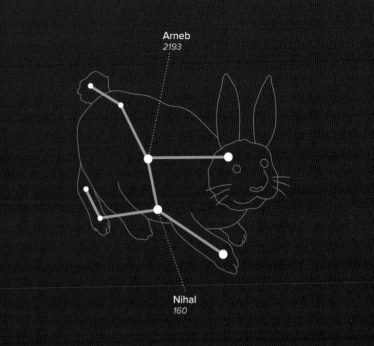

Arneb
2193

Nihal
160

ORION *The Hunter*

BACKGROUND/MYTHOLOGY In my opinion, Orion is the best constellation! It's the centerpiece of the wonderful group of winter constellations that I call "Orion and His Gang." According to Greek mythology, Orion was a mighty nocturnal hunter. In one of the many varying tales about his death, he was killed in a battle with a scorpion sent by the god Apollo. Apollo was angry that Orion and his sister Artemis, the goddess of the moon, had fallen in love. After his death, Artemis had his body placed in the stars.

OBSERVATION NOTES The three stars that make up Orion's belt are the constellation's hallmark. Rigel is the constellation's brightest star and marks Orion's left knee. Orion's second-brightest star, Betelgeuse, represents the armpit of the mighty hunter. Betelgeuse is a supergiant star that regularly swells out to nearly a billion miles in diameter. Below Orion's belt are three more stars in a row that make up the hunter's sheath. The middle "star" is fuzzy to the naked eye because it's actually a huge cloud of hydrogen gas where stars are being born, called a nebula. Technically referred to as M42, it's a must-see with a small telescope; you'll see four newborn stars arranged in a trapezoid. These stars are so bright that they light up the surrounding hydrogen gas like a neon light.

LEPUS *The Rabbit*

BACKGROUND/MYTHOLOGY There are many mythological stories about Lepus. The one I like best is the story about how Lepus is not only elusive but constantly harasses and plays dirty tricks on Orion, the Hunter.

OBSERVATION NOTES It's pronounced Lee-pus—what a great name for a leaping celestial rabbit. Now, I have to admit that Lepus certainly won't blow you away with its brilliance. It's small, and it's not all that bright, especially compared with how massive and brilliant Orion is. The truth is most constellations really don't look all that much like what they're supposed to be. Ancient civilizations used them as rough visual aids to help tell their stories. Back then, stories were told by word of mouth, and without any light pollution, you could really see the stars, making it a little easier to stretch the imagination enough to spot constellations like this somewhat random arrangement of stars. Next door, to the left of Lepus, is the constellation Canis Major, Orion's big hunting dog. It really resembles what it's supposed to, and you can see the big canine in hot pursuit of Lepus, even though you can't see Lepus all that well.

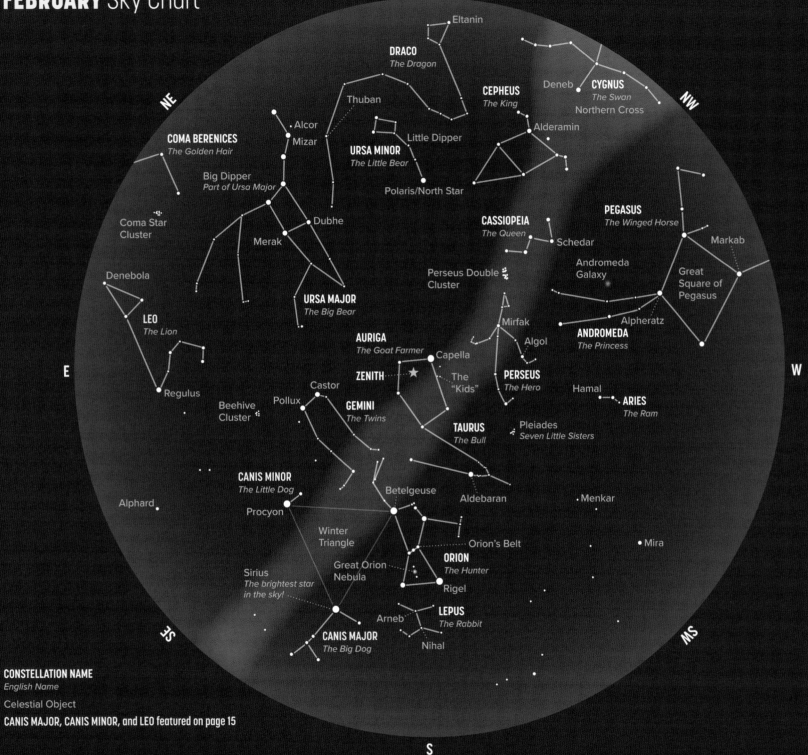

FEBRUARY Sky Chart

N

Eltanin

DRACO
The Dragon

Thuban

CEPHEUS
The King

Deneb

CYGNUS
The Swan
Northern Cross

NE

NW

Alcor

Mizar

COMA BERENICES
The Golden Hair

Little Dipper

URSA MINOR
The Little Bear

Alderamin

Big Dipper
Part of Ursa Major

Polaris/North Star

PEGASUS
The Winged Horse

Coma Star
Cluster

Dubhe

CASSIOPEIA
The Queen

Schedar

Markab

Merak

Andromeda
Galaxy

Great
Square of
Pegasus

Denebola

Perseus Double
Cluster

URSA MAJOR
The Big Bear

Mirfak

Alpheratz

LEO
The Lion

ANDROMEDA
The Princess

Regulus

AURIGA
The Goat Farmer

Capella

Algol

PERSEUS
The Hero

Hamal

ARIES
The Ram

E

ZENITH

The
"Kids"

W

Beehive
Cluster

Castor

Pollux

GEMINI
The Twins

TAURUS
The Bull

Pleiades
Seven Little Sisters

Alphard

CANIS MINOR
The Little Dog

Betelgeuse

Aldebaran

Menkar

Procyon

Winter
Triangle

Orion's Belt

Mira

Great Orion
Nebula

ORION
The Hunter

Sirius
*The brightest star
in the sky!*

Rigel

LEPUS
The Rabbit

Arneb

CANIS MAJOR
The Big Dog

Nihal

SE

MS

S

CONSTELLATION NAME
English Name

Celestial Object

CANIS MAJOR, CANIS MINOR, and LEO featured on page 15

Sky chart is representative of 8 p.m. in February, 12 a.m. in December, and 5 a.m. in October.

February night skies are frantic with bright stars and constellations. The great winter constellations are in prime time, and you don't want to miss the great show. Turn off the tube, get off that couch, and get outside! You'll be glad you did!

February night skies are fantastic. In my opinion, February gifts you with some of the best stargazing of the year. It's not the most comfortable weather, but wonderful celestial jewels await you in the evening heavens, so button up that overcoat, get a thermos with a warm drink, and prepare to be dazzled. If you're not already lucky enough to be out in the countryside, away from city lights, try to get out there, but even if you're restricted to urban stargazing you'll be delighted by the starlight. More than half of the brightest stars you see throughout the course of the year light up our February celestial stage.

The main act is in the southern and southeastern sky. Face that way, and you'll get an eyeful of what I call "Orion and His Gang." I just love, love, love this part of the heavens, especially the main guy, the majestic constellation Orion, the Hunter. Three distinctive stars outline Orion's belt, and he is also home to the bright stars Rigel and Betelgeuse. Also here, you'll find Taurus, the Bull; Auriga, the Chariot Driver; Gemini, the Twins; Lepus, the Rabbit; Canis Minor, the Little Dog; and Canis Major, the Big Dog, with the bright star Sirius representing the nose of the Canis Major. Sirius is the brightest star in the night sky at any time of the year. The name Sirius is of Greek origin and translates to "the scorcher."

All the stars you see in our sky are part of our home Milky Way Galaxy, which consists of a bright nucleus of stars connected to several spiral arms of stars. Our sun is located in the middle of one of the spiral arms. Given the logjam of bright stars and constellations that you see in the early evening winter sky, you'd think that you were facing toward the center of our galaxy, but you're actually facing away from the center of the Milky Way. Thankfully, you luck out as you're right on top of one of the brighter neighboring spiral arms, which astronomers call the Orion Arm. Without a doubt, the summertime stars are wonderful but nowhere near as dazzling as the winter constellations, and there are no mosquitoes!

One of the truly amazing sights within Orion's posse is the Winter Triangle, which is made up of three bright stars from three separate constellations. Betelgeuse, which represents the right armpit of Orion, is the star at the upper right-hand corner of the triangle. Sirius, the brightest star in Canis Major, marks the lower corner of the triangle, and the upper left corner is occupied by Procyon, the brightest star in Canis Minor, aka the Little Dog. Like the three stars that make up Orion's belt, the stars of the Winter Triangle have nothing to do with each other astronomically. They just happen to be arranged in a perfect triangle from our vantage point on Earth. In several thousand years, these stars will have moved enough that the Winter Triangle will become lopsided, but the triangle will no doubt remain one of the great wonders of the night sky for generations to come.

Another spectacular celestial wonder in the night sky right now is the Great Orion Nebula, a gargantuan stellar factory. I guarantee that when you point your telescope at it, even a smaller scope or binoculars, you'll be awed when you look at it for the first time. Even to the naked eye it's impressive looking.

To find that nebula, look below the three bright stars in a row that make up Orion's belt. Just to the lower left of the belt, you'll see a row of three fainter stars lined up diagonally that make up Orion's sheath.

Right away you can't help but notice that the middle "star" of the sheath is fuzzy. That's it! You've arrived at Orion Nebula, which is around 1,500 light-years away. (Remember, just 1 light-year equals almost 6 trillion miles!)

Before you aim your binoculars or your new Christmas telescope at it, I want you to do something. I want you to hold out your hand at arm's length so you can easily cover up the Orion Nebula with the tip of your thumb. That thumb of yours is covering a giant cloud of hydrogen gas over 30 light-years in diameter. That's almost 180 trillion miles across, or about 20,000 times the diameter of our solar system. Your thumb is covering up all that!

Okay, now aim your telescope at it. After initially taking it in you may notice that the nebula has a greenish tinge to it. Another feature you'll observe, even with the smallest of telescopes, is a lopsided trapezoid composed of four stars. If your telescope is larger, you may see a fifth or even a sixth star in the trapezoid, which is referred to as Trapezium. Galileo Galilei was the first known person to document the Trapezium, and he first saw it on February 4, 1617. The four stars that make up the Trapezium are stellar infants, probably less than 50,000 years old, with at least one that may only be 10,000 years young. They're huge stars, more than 15 times the mass of our sun, and the stars are only separated by about 1.5 light-years. One of the stars is estimated to have a surface temperature over 70,000 degrees Fahrenheit, more than six times the surface temperature of the sun. All of the heat and radiation pouring out of the stars in that area of space cause the surrounding hydrogen gas to glow like a giant fluorescent light. Astronomers refer to this kind of nebula as an emission nebula.

The stars of the Trapezium were born out of the Orion Nebula, and the Orion Nebula isn't finished making stars. In fact, that giant cloud of gas could produce many more stars in the future, maybe another 10,000 stars the size of our sun or larger.

The Hubble Space Telescope, which is far more powerful than backyard telescopes, has even detected developing solar systems around some of the stars of the Orion Nebula, but these potential planets may not survive. Stellar winds gusting over 5 million miles per hour are constantly destroying developing planet families. In fact, tremendous stellar currents from several stars can collide to cause a perfect cosmic storm, otherwise known as complete celestial chaos!

In the eastern February evening sky, there is a sign of spring with the first appearance of the constellation Leo, the Lion. The right side of Leo resembles a backward question mark in the eastern sky and, unlike many constellations, actually bears a passing resemblance to what it's supposed to depict. Without too much trouble, you can see how the backward question mark outlines the profile of a lion's head. You'll get a lot better look at it later in the evening as it rises higher. To the lower left of Regulus, Leo's brightest star, are three stars that form a triangle marking the backside of the dangerous lion.

Just below Leo's rear end is what's known as the Leo Triplet, a trio of galaxies around 35 million light-years away. Astronomically they're known as M65, M66, and NGC 3628. They may be seen with a small to moderate telescope, but it is challenging. By the way, when using a telescope in winter, remember to let your telescope sit outside for at least half an hour so the optics adapt to the cold winter air.

MONKEY HEAD NEBULA

SEAGULL NEBULA

HEART NEBULA (right)

FEBRUARY Featured Constellations

Numbers under star names represent light-years

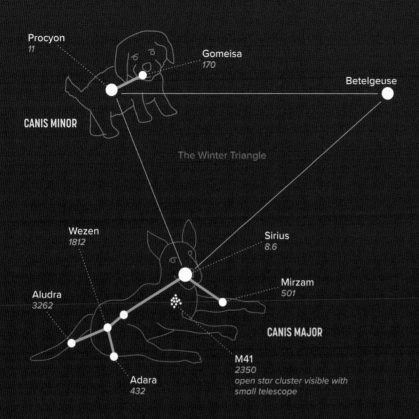

Procyon
11

Gomeisa
170

Betelgeuse

CANIS MINOR

The Winter Triangle

Wezen
1812

Sirius
8.6

Aludra
3262

Mirzam
501

CANIS MAJOR

Adara
432

M41
2350
*open star cluster visible with
small telescope*

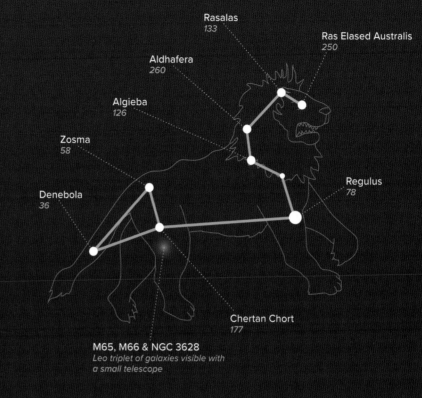

Rasalas
133

Ras Elased Australis
250

Aldhafera
260

Algieba
126

Zosma
58

Regulus
78

Denebola
36

Chertan Chort
177

M65, M66 & NGC 3628
*Leo triplet of galaxies visible with
a small telescope*

CANIS MAJOR/CANIS MINOR *The Big Dog/The Little Dog*

BACKGROUND/MYTHOLOGY The mighty Orion's hunting dogs, Canis Major and Minor (Latin for the big and little dogs, respectively), faithfully follow along behind the great hunter.

OBSERVATION NOTES Canis Minor really isn't a very large constellation. About all there is to it is the bright star Procyon and the dimmer star Gomeisa. On the other hand, Canis Major really looks like a hunting dog. You can easily see a big dog standing up on its hind legs, raring to go. At Canis Major's nose is Sirius, the brightest star in the night sky. Its brightness is mainly due to its proximity to our solar system; at 8.6 light-years, it's about 50 trillion miles away. Believe it or not, that's considered close in astronomy. At the other end of Canis Major, there's a truly amazing star, Aludra. It's well over 3,000 light-years away, and yet you can see it easily. Given its distance, it has to be super bright for you to see it, and it really is, kicking out over 100,000 times as much light as our sun.

LEO *The Lion*

BACKGROUND/MYTHOLOGY In Greek mythology, Leo is a huge lion that terrorized the countryside. Its hide was so thick, spears and arrows couldn't pierce it. However, the mighty hero Hercules wrestled the beast with his bare hands, grabbing one of its paws and slitting Leo's throat with its own giant feline claw.

OBSERVATION NOTES While the constellation Orion, the Hunter, is without a doubt the king of the winter skies, Leo is the kingpin of the spring constellations. It's not as flashy as Orion, but it's very distinctive, flying in the southeastern evening sky. The right side of it looks just like a backward question mark. The period at the bottom is Regulus, the brightest star in Leo. At almost 78 light-years away, the light you see from it left near the end of World War II. The left side of Leo is a triangle of moderately bright stars that outline the lion's hindquarters and tail. Just below Leo's rear end is what's known as the Leo Triplet, a trio of galaxies around 35 million light-years away. Astronomically, they're known as M65, M66, and NGC 3628. They may be seen with a small to moderate telescope, but it is challenging.

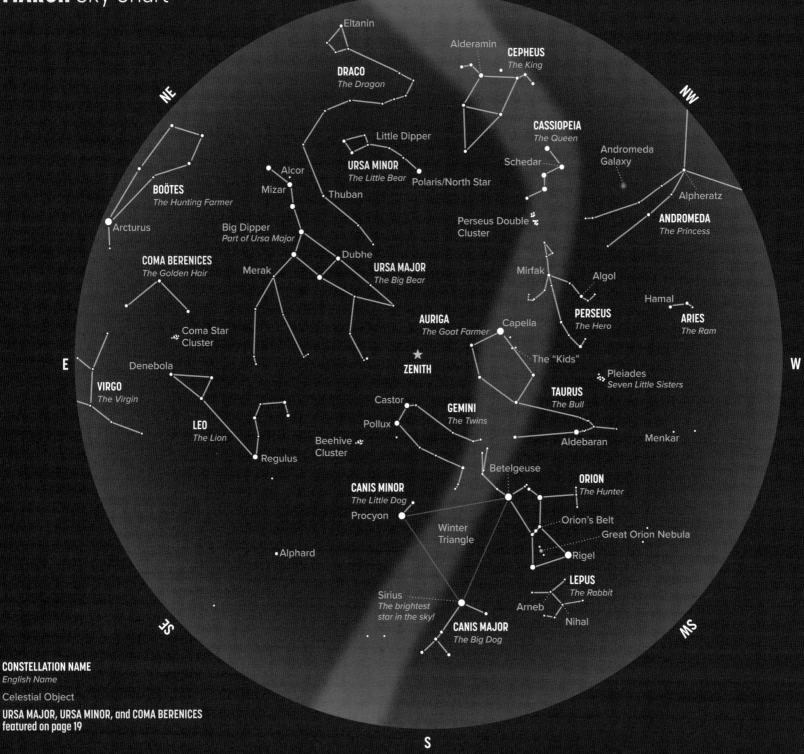

MARCH Sky Chart

N

Eltanin

Alderamin

CEPHEUS
The King

NE

DRACO
The Dragon

NW

Little Dipper

CASSIOPEIA
The Queen

Schedar

Andromeda
Galaxy

Alcor

URSA MINOR
The Little Bear

Polaris/North Star

Alpheratz

Mizar

Thuban

ANDROMEDA
The Princess

BOÖTES
The Hunting Farmer

Arcturus

Big Dipper
Part of Ursa Major

Perseus Double
Cluster

COMA BERENICES
The Golden Hair

Dubhe

Mirfak

Algol

Hamal

Merak

URSA MAJOR
The Big Bear

PERSEUS
The Hero

ARIES
The Ram

Coma Star
Cluster

AURIGA
The Goat Farmer

Capella

E

★
ZENITH

The "Kids"

W

Denebola

Pleiades
Seven Little Sisters

VIRGO
The Virgin

Castor

GEMINI
The Twins

TAURUS
The Bull

LEO
The Lion

Pollux

Beehive
Cluster

Aldebaran

Menkar

Regulus

Betelgeuse

CANIS MINOR
The Little Dog

ORION
The Hunter

Procyon

Winter
Triangle

Orion's Belt

Great Orion Nebula

Alphard

Rigel

LEPUS
The Rabbit

Sirius
*The brightest
star in the sky!*

Arneb

Nihal

SE

CANIS MAJOR
The Big Dog

SW

S

CONSTELLATION NAME
English Name

Celestial Object

URSA MAJOR, URSA MINOR, and COMA BERENICES
featured on page 19

Sky chart is representative of 9 p.m. in March, 12 a.m. in January, and 4 a.m. in November.

It's the best of all worlds when you look out from Earth this month. March stargazing is fantastic because you still have Orion and all of the great constellations of winter, but on most nights the chill of winter has eased a bit. In fact, spring begins this month!

The evenings are starting later now, and the prime winter constellations are found a little more to the west each evening. Unfortunately, the eastern skies on March evenings are not as action-packed and don't have as many bright constellations, at least to the naked eye. One exception is Leo, the Lion, which is prowling higher and higher into the night sky as March stalks toward April.

Leo is one of the 13 constellations of the zodiac; the zodiac traces a band across all of the celestial sphere. The middle of that band is called the ecliptic, the path that the sun seems to take across the background of stars as the Earth orbits the sun every year. The zodiac band and the ecliptic aren't a straight line but curve across the heavens like a wave, a reflection of the fact that the Earth's axis is tilted 23.5 degrees as it orbits around the sun. It's no coincidence that 12 of the signs of the zodiac (or horoscope) bear the same names as the famous astrological signs, as these constellations once served as a type of calendar. Ancient astronomers took this a step further; the practice of astrology is an attempt to predict the future based on the movement of the moon and planets among the zodiac constellations. How that's supposed to tell you if you're going to meet the love of your life or get hit by a truck is beyond me!

Even though astrology might not be good at predicting the future, being able to spot the zodiac band is useful for another reason: Earth's moon and the other planets in the solar system all orbit around the sun in nearly the same plane. So not only does the sun travel along the ecliptic and zodiacal band, so do the moon and most of the other planets. The only planet that strays out of the zodiacal band is Pluto, and it is now officially considered a dwarf planet.

A group of constellations in the northern sky are always above the horizon; such constellations are called circumpolar. Two of them are Ursa Major, the Big Bear, and Ursa Minor, the Little Bear. Even though it's visible in the night sky over most of the United States throughout the year, the Big Bear is now higher in the sky and a lot easier to see in the early evening. Most people have never seen Ursa Major in its entirety, but everybody's seen the Big Dipper, which is the brightest part of the Big Bear and outlines the rear end and tail of one of the biggest constellations in the heavens.

Even though the Big Dipper is only part of a constellation and not one of the 88 officially recognized constellations, it can help you begin to understand the movements of the celestial theater and how the stars move across our sky every night (and day for that matter).

In the evening, if you face roughly north just after twilight, you'll see the Big Dipper standing more or less diagonally on its handle in the northeastern sky. Unless you're watching from areas with the worst light pollution, you should have no trouble seeing it. You can use the two stars in the pot section of the Big Dipper, Dubhe and Merak, to find Polaris, otherwise known as the North Star. These two stars are opposite the dipper's handle. Just draw a line from Merak to Dubhe in your mind's eye and continue that line down and to the left, and you'll run into Polaris. That line from Merak and Dubhe doesn't point exactly at Polaris, but it's close enough. If you extend your clenched fist at arm's length, three fist-widths should get you from Dubhe to the North Star.

Polaris is certainly not the brightest star in the sky, but it is an important one. I exclusively call it the linchpin of the night sky because everything in the northern heavens completes one circle around the North Star every 24 hours. That's because the North Star is shining almost directly above the Earth's North Pole. If you were standing on the North Pole, Polaris would be directly overhead, at the zenith, and everything in the northern skies would circle the pole star every night and day. Stars close to Polaris make tight little circles, and stars closer to the horizon make progressively larger circles, but all stars make one complete circuit in 24 hours, a reflection of the time it takes Earth to complete one full rotation.

We don't live at the North Pole, although it feels like that at times in winter! Most of the United States lies somewhere between ⅓ to ½ of the distance from the equator to the North Pole. The rule of thumb is that Polaris's altitude above the northern horizon is equal to your position in latitude, so if you're at 45 degrees north latitude, the North Star will be 45 degrees above the horizon. Given the range of latitudes in the US, the North Star's altitude depends on where you are; it is usually found between 30–45 degrees above the northern horizon.

Stars and constellations in the northern sky make very small circles around Polaris and never get below our horizon. They're called circumpolar stars, and examples include the stars of the Little Dipper and most or all of the stars in the Big Dipper, again depending on where you live. In far southern states, the Big Dipper's handle does briefly dip below the horizon. Stars farther away from Polaris in the celestial dome make much larger circles around the North Star and dip below the northern horizon. This makes those stars appear to rise in the east and set in the west, just like the sun and moon.

When it comes to the constellation itself, the North Star is the brightest star of the Little Dipper and shines at the end of its handle. The Little Dipper is nearly upright below the Big Dipper and is a lot tougher to see, especially in areas with city lights. To see it, look for two moderately bright stars to the right of Polaris. They're Kochab and Pherkad, and they appear on the opposite side of the Little Dipper's pot section. The other two stars in the pot and the other two handle stars can be tricky to see unless you're in the country, away from light pollution.

The seven stars that make up the Little Dipper also outline the figure the constellation is supposed to depict: the Little Bear, otherwise known as Ursa Minor. Polaris marks the end of the Little Bear's tail. Seeing a little bear when you look at the Little Dipper requires your imagination to shift into a somewhat higher gear.

Ursa Major looks a little bit more like a bear. Look just to the upper left of the pot section for three dimmer stars that form a skinny triangle that outlines the Big Bear's head. From that skinny triangle, look to the upper right for two stars right next to each other that should be obvious. Those stars are called Talitha and Al Kaprah, and they mark the bear's front paw. Between the front paw and the triangular head is a star that represents the bear's knee, and once you spot that, you've seen one of the front legs of Ursa Major. Unfortunately, there are no stars that make up the other front leg, so don't feel bad if you can't find it. Two curved lines of stars outline the bear's back legs, but the one in the foreground is much easier to see. Just look to the lower right of the two front paw stars for two more stars right next to each other. That's Tania Borealis and Tania Australis, and they make up one of the Big Bear's back paws. In order to see the rest of the rear leg, look for two stars to the lower left of the rear paw stars; these stars join the pot of the Big Dipper and represent the rear end of the Big Bear. That's it: You've just seen one of the largest constellations!

WHIRLPOOL GALAXY (right)

RUNNING MAN NEBULA   **LEO TRIPLET GALAXIES**

Numbers under star names represent light-years

 URSA MAJOR

Talitha
48

Al Kaprah
423

Tania Borealis
135

Muscida
184

Tania Australis
249

M81
12 million
Bode's Galaxy

M82
12 million
Cigar Galaxy

Merak
79

Dubhe
124

Phecda
84

Three fist-widths at arm's
length to Polaris (North Star)

Megrez
81

Kochab
127

Alioth
81

Mizar
78

Polaris
432
The North Star

URSA MINOR

Pherkad
423

Alcor
81

Alkaid
101

Coma Star Cluster
250
a bright open cluster of young
stars that can be seen with a
small telescope

M64
20 million
"Blackeye" Galaxy

Diadem
47

URSA MAJOR/URSA MINOR *The Big Bear/The Little Bear*

BACKGROUND/MYTHOLOGY Zeus fell in love with Callisto, a hunting companion of the goddess Artemis. In revenge, Hera, Zeus's wife and the queen of the gods, turned Callisto into a big bear. Arcus, Callisto's son, grew up to be a hunter and almost shot his mother, not knowing who the bear really was. Zeus prevented this by turning Arcus into a little bear. Hera was still very angry, so Zeus hurled Callisto and Arcus into the stars so they would never be harmed.

OBSERVATION NOTES Ursa Major and Ursa Minor are the Latin names for two of the most famous constellations, the Big Bear and Little Bear. The Big Bear is one of the largest constellations in the heavens. Most people have only seen the portion of it known as the Big Dipper. In the spring evening sky, the Big Bear is flying upside down in the high northern sky. Below the Big Dipper is the upright Little Dipper, which doubles as Ursa Minor, the Little Bear. It looks much more like a dipper than a baby bear. At the end of the handle of the Little Dipper is Polaris, the North Star. Polaris isn't the brightest star, but it's what I call the linchpin of the sky. All the stars in our sky make one complete circle around Polaris every day. That's because the North Star shines directly above Earth's north pole.

COMA BERENICES *The Golden Hair*

BACKGROUND/MYTHOLOGY Coma Berenices is actually a relatively new constellation. Created in the early 17th century, its "myth" is actually based on a true story. Queen Berenice was the wife of the famous Egyptian Pharaoh Ptolemy III. The great pharaoh led his troops into battle, and his wife promised the gods that she would cut off her beautiful locks of hair if her husband returned safe and sound. Her sheared hair allegedly found its way to the heavens.

OBSERVATION NOTES Spring is the stargazing equivalent of the doldrums. The constellations are mainly dim and dull, and Coma Berenices is no exception. Its one redeeming value is that it resembles what it's supposed to represent: long locks of flowing hair being tossed in a breeze. Coma Berenices is more of a star cluster than a constellation. Found in the southeastern sky, it looks like faint strands of hair. You honestly really won't see it all that well in the lights of the cities and suburbs. You really need to be out in the countryside. The actual star cluster is a congregation of young stars about 250 light-years away. Believe it or not, even at 1,400 trillion miles, it's actually one of the closer young clusters.

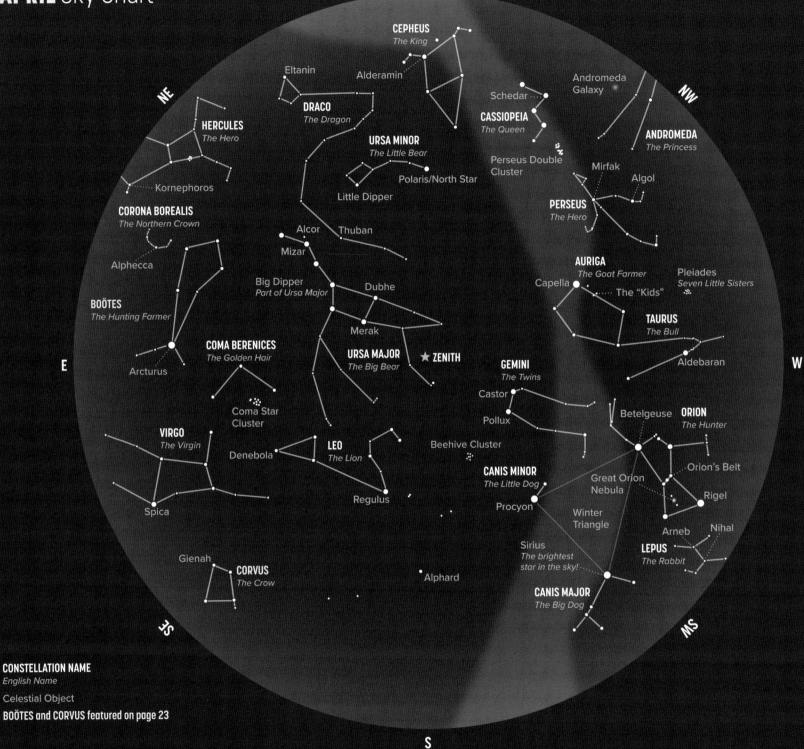

APRIL Sky Chart

N

CEPHEUS
The King

Eltanin

Alderamin

DRACO
The Dragon

Schedar

Andromeda
Galaxy

NW

NE

HERCULES
The Hero

URSA MINOR
The Little Bear

CASSIOPEIA
The Queen

ANDROMEDA
The Princess

Polaris/North Star

Perseus Double
Cluster

Mirfak

Algol

Kornephoros

Little Dipper

CORONA BOREALIS
The Northern Crown

PERSEUS
The Hero

Alcor

Thuban

AURIGA
The Goat Farmer

Pleiades
Seven Little Sisters

Alphecca

Mizar

Capella

The "Kids"

Big Dipper
Part of Ursa Major

Dubhe

BOÖTES
The Hunting Farmer

TAURUS
The Bull

Merak

COMA BERENICES
The Golden Hair

URSA MAJOR
The Big Bear

★ ZENITH

E

Arcturus

Aldebaran

W

GEMINI
The Twins

Castor

ORION
The Hunter

Betelgeuse

Coma Star
Cluster

Pollux

VIRGO
The Virgin

Beehive Cluster

Orion's Belt

Denebola

LEO
The Lion

Great Orion
Nebula

Rigel

CANIS MINOR
The Little Dog

Spica

Regulus

Winter
Triangle

Arneb

Nihal

Procyon

Gienah

CORVUS
The Crow

Sirius
*The brightest
star in the sky!*

LEPUS
The Rabbit

Alphard

CANIS MAJOR
The Big Dog

SE

SW

S

CONSTELLATION NAME
English Name

Celestial Object

BOÖTES and **CORVUS** featured on page 23

Sky chart is representative of 9 p.m. in April, 12 a.m. in February, and 4 a.m. in December.

Say goodbye to the great constellations of winter, like Orion and his surrounding cast of characters, but say hello to more comfortable stargazing!

So, you've arrived at April. The nights are growing warmer, and you can take off at least some of your layers while stargazing, but there is a trade-off here, actually a couple of trade-offs. For one thing, while the nights are getting warmer, they are also getting shorter. Now, it's not dark enough for decent stargazing until well after 8 p.m. That makes it more difficult for amateur astronomers, especially those who have an early wake-up call to get to work. Another trade-off is that the best and brightest stars and constellations of the year, at least in my opinion, are starting to head for the celestial exits. The mighty constellation Orion, the Hunter, and his gang of bright constellations are a little farther to the west every night at the start of evening. The Earth, in its perpetual orbit around the sun, is gradually turning away from the great stars of winter. By mid to late May, they will pretty much be gone until fall when they show up in the eastern heavens. That's OK, though, because it's fun to have different constellations in the evening sky. Variety is the spice of life—and stargazing—and there are still many treasures in the rest of the heavens. In the high southeastern sky, look for a backward question mark that outlines the heart, chest, and head of Leo, the Lion. The semi-bright star Regulus, Leo's heart, marks the dot at the bottom of the question mark. To the lower left of Leo's head is a small but distinct triangle that makes up the great feline's rear and tail.

In the low southeastern sky, one of the less remarkable constellations, Corvus, the Crow, is on the rise. If you look at that lopsided trapezoid of stars and can see a crow, you have a better imagination than I do. Seriously, when it comes to constellations,

ancient cultures used them as rough visual aids to tell their stories. They didn't really care that the constellations didn't look much like what they were supposed to be.

In the low eastern sky is Boötes, the Hunting Farmer, arguably one of the first constellations of summer, although some insist that it's a spring star pattern. I contend that it's a summer constellation since it stays out in the sky all summer long, even into October.

Like many other stars and constellations, Boötes has several pronunciations. Most people say boot-tees, but I've also heard it pronounced boo-oat-tees; but no matter how you pronounce it, Boötes resembles a giant kite flying on its side in the eastern sky. Finding Boötes in the early evening is easy. Just look for the brightest star you can see in the east. That's Arcturus, the brightest star in Boötes and the second-brightest star in most of the US; if you live in the southern tier of this great country, it's the third-brightest star behind Canopus, which is seen in the extreme southern sky in the winter. No matter where you live, Arcturus is the "star of summer."

If you need rock-solid confirmation you're seeing Arcturus, use the old stargazing rule "arc to Arcturus." Look at the nearby Big Dipper and follow the curve or arc of its handle beyond the end of the handle and you'll run right into Arcturus, which is at the tail of the giant sideways kite. To see the rest of Boötes, just look to the left of Arcturus and without too much trouble you should see the rest of the kite.

Arcturus has a distinctive orange glow to it, typical of stars classified as red giants. Even though Arcturus is 25 times the diameter of our sun, it only has 1.5 times as much mass. Arcturus is running out of hydrogen fuel in its core. When that happens, stars puff out into red giants. This will happen to our own

sun in about 5 to 6 billion years, so when you see Arcturus, you're looking at our future. Arcturus is just about 37 light-years away (about 214 trillion miles), and believe it or not, that's considered a nearby star. It took 37 years for the light from that star to reach us. Just think of what you were doing and where you were living 37 years ago (assuming you were alive)—that's when the light you're seeing left Arcturus. Any time you gaze at Arcturus, or any other star, you're looking back in time.

Arcturus hasn't always been as close to our solar system or as bright as it is now. Actually, all the stars in our sky are zipping along at different speeds and in different directions as they orbit around the center of our galaxy. It certainly doesn't seem like that because you see the same constellations that civilizations have witnessed for thousands of years. That's because the stars are so far away that even at their celestial breakneck speeds, the star patterns pretty much stay the same for thousands of years.

Arcturus, in particular, is most definitely in the express lane on the interstellar speedway. For millions of years, Arcturus has been racing toward the general direction of our solar system at an incredible 90 miles per second! According to many astronomers, Arcturus wasn't even visible in the night sky 500 million years ago (not long ago by astronomical standards). At 37 light-years, it's about as close as it'll get toward Earth, as it's now beginning to pull away. So, take a good look now because about 500 million years from now, Arcturus will fade into oblivion.

As with many constellations, there are many mythological stories about how Boötes got in the sky. In the best-known tale, a desperately impoverished Boötes invents a plow that can be pulled by oxen. Demeter, the goddess of agriculture, was so impressed with Boötes that when he died, she transformed his body into a constellation. While he was alive, one of Boötes' passions was hunting, so when he died, Demeter placed Boötes in the heavens and put him on an everlasting pursuit of the Big Bear, Ursa Major.

That's a nice story, but the one I love involves one of my heroes, Bacchus, the Roman god of wine. Boötes, also known as Icarius, was the proprietor of a large vineyard and grew the best grapes in all the land. Bacchus was so impressed with Boötes' vineyard that he revealed the secret of winemaking to him. Boötes immediately brought his friends together for a wine-tasting party that soon got out of hand. Most of the guests woke up the next day feeling ill. Not knowing about the intoxicating effects of wine, many of them thought that Boötes was trying to poison them. Before the first winemaker woke, his former friends took spears and rocks to him. When Bacchus heard this, he took pity on Boötes and transformed his lifeless body into a constellation. So, the next time you're out there in the evening after dark, raise your glasses to the constellation Boötes!

PINWHEEL GALAXY

VIRGO CLUSTER OF GALAXIES

SEAGULL NEBULA (right)

Numbers under star names represent light-years

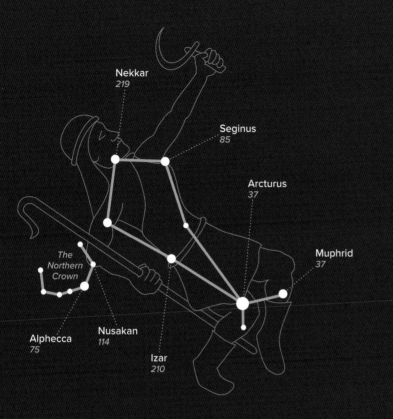

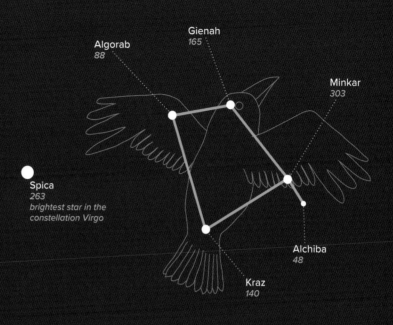

BOÖTES *The Hunting Farmer*

BACKGROUND/MYTHOLOGY According to Greek mythology, Boötes was the son of Demeter, the goddess of agriculture. His father was a mortal. As a result, Boötes was half-god. As a baby, he was given to a mortal family to be raised as a common farmer. When he had grown into an adult, his parents were killed in a tragic accident, and his brother ran away with the money they left behind. Left to farm the land, Boötes invented the first ox-driven plow. He became very wealthy, and to honor him, the gods placed his image among the stars.

OBSERVATION NOTES Boötes, the hunter and farmer, actually looks a lot like a giant kite rising in the eastern sky. That kite has a very bright star—called Arcturus—at its tail. If you need rock-solid confirmation you're seeing Arcturus, use the old stargazing trick "arc to Arcturus." Follow the curve or arc of the Big Dipper's handle beyond the end of the handle and you'll run right into Arcturus. Not only is Arcturus the brightest star in Boötes, it's also the fourth-brightest star in the sky. A red giant about 21 million miles in diameter, Arcturus has an orange tinge even when seen with the naked eye.

CORVUS *The Crow*

BACKGROUND/MYTHOLOGY While Corvus isn't much for looks, it has quite a story. According to legend, crows were not always the ugly black birds we know today. They were intelligent, gorgeous white birds with gold trim, and they sang beautifully. They were faithful servants of the gods. That all changed when Corvus, the Crow, erred when he was sent by Apollo to fetch water from a magical fountain. When his task began, he promptly got lost, and to make matters worse, he stopped for a drink and imbibed a little too much. Afterward, he tried once again to find the fountain, failed miserably, and when Corvus flew back, Apollo was furious and took out his anger on all crows. He turned them all into ugly black birds and replaced their beautiful song with an irritating caw.

OBSERVATION NOTES Corvus is not one of those constellations that cause you to gaze in awe. The truth is that it's somewhat unimpressive and not all that bright, but it's pretty easy to find. Flying low in the southeastern evening sky just above the horizon, Corvus resembles a lopsided diamond.

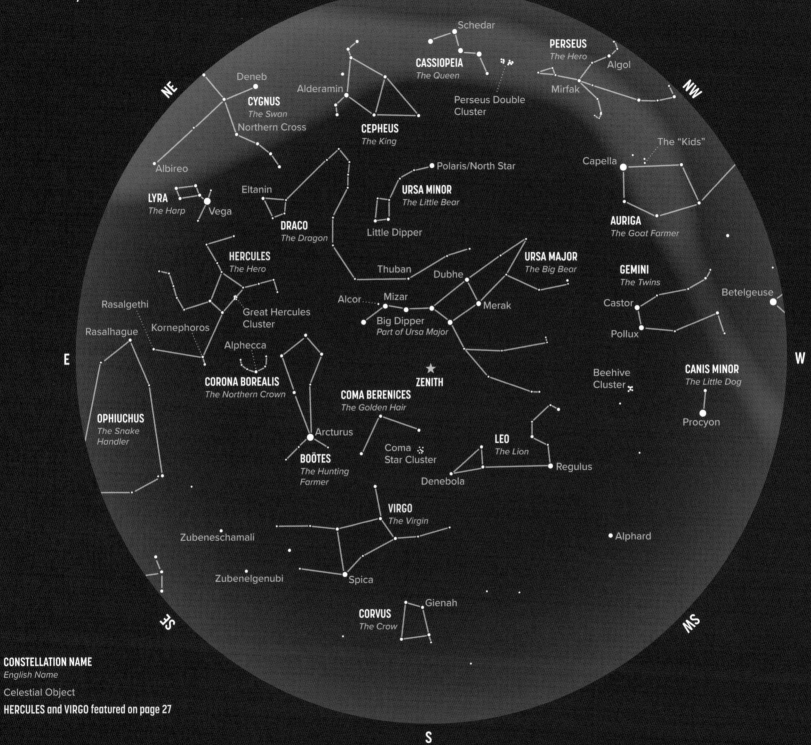

MAY Sky Chart

N

Schedar

PERSEUS
The Hero

CASSIOPEIA
The Queen

Algol

NE

Deneb

Alderamin

Mirfak

NW

CYGNUS
The Swan
Northern Cross

CEPHEUS
The King

Perseus Double
Cluster

The "Kids"

Albireo

Polaris/North Star

Capella

LYRA
The Harp

Eltanin

URSA MINOR
The Little Bear

Vega

AURIGA
The Goat Farmer

Little Dipper

DRACO
The Dragon

HERCULES
The Hero

Thuban

Dubhe

URSA MAJOR
The Big Bear

GEMINI
The Twins

Betelgeuse

Rasalgethi

Great Hercules
Cluster

Alcor

Mizar

Merak

Castor

Rasalhague

Kornephoros

Big Dipper
Part of Ursa Major

Pollux

Beehive
Cluster

E

Alphecca

ZENITH

CANIS MINOR
The Little Dog

CORONA BOREALIS
The Northern Crown

COMA BERENICES
The Golden Hair

W

OPHIUCHUS
*The Snake
Handler*

Arcturus

Coma
Star Cluster

LEO
The Lion

Procyon

BOÖTES
*The Hunting
Farmer*

Regulus

Denebola

VIRGO
The Virgin

Zubeneschamali

Alphard

Zubenelgenubi

Spica

Gienah

SE

CORVUS
The Crow

MS

S

CONSTELLATION NAME
English Name

Celestial Object

HERCULES and VIRGO featured on page 27

Sky chart is representative of 10 p.m. in May, 2 a.m. in March, and 5 a.m. in January.

The nights are warmer but shorter, with darkness falling after 10 p.m. Catch an afternoon nap so you can stargaze in comfort. The great winter constellations are saying goodbye in the western sky, and the summer shiners are on the rise in the east.

It's a definite surprise when it comes to watching what's left of the brilliant winter constellations. They're crashing in the west, and the night sky has changed! As the Earth continues its annual journey around the sun, you're turning away from the direction of space occupied by Orion and company, and you are now pointing in the direction of the noticeably less brilliant constellations of springtime.

There's still time to check out at least what's left of the winter constellations in the west, but their days, or should I say nights, are definitely numbered. When it finally gets dark enough (around 9:30 p.m. this time of year), Orion is already partially set in the west. Hovering above the horizon, you can still barely see the three bright stars in a row that outline the great hunter's belt. Above the belt is the bright star Betelgeuse, which represents the armpit of Orion.

To be truly honest with you, many amateur astronomers, including this star watcher, agree that until the summer constellations, such as Cygnus and Scorpius, make their appearance, you are officially in the spring doldrums of evening stargazing. Even so, it's still worth your time to make the stars your old friends. For one thing, in May, it's a heck of a lot more comfortable out there than in winter, and in most places, the mosquitoes aren't at peak numbers yet.

The semibright but distinctive constellation Leo, the Lion, is just beginning to head into the western half of the sky while Boötes, the kite-shaped farmer, is busy hunting the Big Bear (Ursa Major) and climbing higher and higher in the eastern half of the celestial dome. Speaking of Ursa Major, the Big Dipper, which represents the rear and tail of the Big Bear, is almost upside down and nearly overhead in the high northern sky. The Big Dipper is always upside down in the evening this time of year and according to old American folklore, that's why there is so much rain in the spring.

In the lower eastern sky, Hercules, the Hero, is on the rise. When it comes to constellations, it certainly is in the lower tier in brightness, but I truly love it for its great celestial treasure, Messier Object 13, or as it's better known, the Great Hercules Cluster. I love to show off this heavenly gem to folks who come to my star parties. You should hear the "wows" and the "holy cows" when they check it out through my large telescope.

When you see the cluster, you're looking at a minimum of 300,000 stars crammed into a tight ball that is less than 150 light-years across and about 25,000 light-years away. This time of year, you can start your search for the great Hercules globular cluster as soon as it gets dark enough, after 10 p.m. For all practical purposes, it's not visible to the naked eye, but with a dark enough sky, such as in the outer suburbs or the countryside, you should be able to hunt it down with a decent pair of binoculars or a small telescope. When you get it in your scope, I can just about guarantee you'll fall in love with one of the true jewels of the heavens, especially if you're observing it in the dark countryside.

At the end of evening twilight, M13 is located on the west side of the trapezoid that makes up the center portion of the constellation Hercules. I think the easiest way to find it is to use the two brightest stars in the late spring sky, Vega and Arcturus. They also happen to be in the eastern half of the May evening sky. Arcturus is the brightest star in the constellation Boötes and the brightest visible in the month of May; it is very high in the east and has a definite orange hue. Vega is about as bright as Arcturus and sports a bluish hue in the low northeastern sky. If you draw a line between Vega and Arcturus, M13 will be just short of the halfway point between them. Scan that area with your binoculars or telescope and see if you can spot what looks like a little fuzzball. Like a lot of worthy celestial treasures, you have to search for it a bit. Think of it as an extra credit project for the beginning stargazer.

The Hercules Cluster is a prime example of a globular star cluster. There are hundreds and hundreds of star clusters in the night sky. Just slowly scan the heavens with any old pair of binoculars, and you can't help but find them. Most of these are what astronomers call open star clusters and are made up of groups of young stars that recently formed out of the same hydrogen gas cloud. The baby stars in these open star clusters are generally anywhere from 50 to 150 million years old, which is hardly any time at all when you're talking astronomy. Globular clusters like M13 are different. They are spherical swarms of thousands or even hundreds of thousands of stars packed in a small area, usually less than 300 light-years in diameter. Globular clusters are made up of old stars that are generally more than 12 billion years old. More than 140 globular clusters form a giant halo around our Milky Way Galaxy. In a way, they are part of the outer structure of our home galaxy, and astronomers refer to them as satellites of the Milky Way because they orbit the galaxy's core. Because of that, globular clusters are a long way away. Again, the Hercules Cluster (M13) is about 25,000 light-years away, and remember, just one light-year equals 6 trillion miles.

M13 is by far the best globular star cluster in the spring and summer sky, but there are several other great ones too. Go ahead and Google "globular clusters," and you'll see what I mean. I do want to point out one other globular cluster that shouldn't be too hard to find. It's also in the constellation Hercules, the Hero. It's M92, and as you can see on the next page, it is really close to M13.

Happy globular hunting!

BEEHIVE STAR CLUSTER

BODES/CIGAR GALAXIES

　GREAT HERCULES CLUSTER (right)

MAY Featured Constellations

Numbers under star names represent light-years

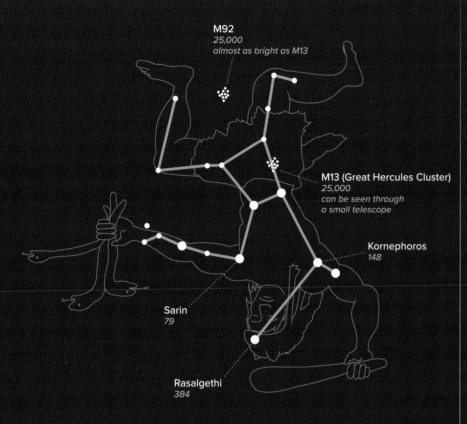

M92
25,000
almost as bright as M13

M13 (Great Hercules Cluster)
25,000
can be seen through
a small telescope

Kornephoros
148

Sarin
79

Rasalgethi
384

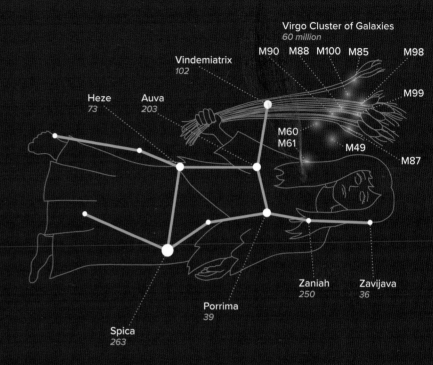

Virgo Cluster of Galaxies
60 million

M90 M88 M100 M85 M98

Vindemiatrix
102

M99

Heze   Auva
73   203

M60
M61   M49

M87

Zaniah
250

Zavijava
36

Porrima
39

Spica
263

HERCULES *The Hero*

BACKGROUND/MYTHOLOGY Hera, the wife of Zeus, cursed Hercules with madness. Hercules killed his wife and child in a moment of insanity and then begged the gods for mercy. They took pity on him but made him atone for his horrible crime by performing 12 acts of heroism. As a reward, his body was placed in the stars.

OBSERVATION NOTES Hercules is one of my favorite summer constellations. While it's not the brightest constellation, it's large, the fifth largest in the sky. Hercules is supposedly pictured in the sky hanging upside down and is visible early in the evening in the low eastern sky. Spotting the constellation is often not easy, especially if there is light pollution, but give it a shot. Even if you don't see the entire constellation, you should be able to spot a trapezoid made up of four moderately bright stars. This is known as the Keystone and outlines Hercules' torso. The best telescope target in the constellation is the Great Hercules Cluster. Even with a smaller telescope, you can see some individual stars at its edges. It's a cluster of at least 300,000 stars about 25,000 light-years away.

VIRGO *The Virgin*

BACKGROUND/MYTHOLOGY Unfortunately, Virgo, the Virgin, is not so prominent in the night sky. It's a large but faint constellation and you really need to be away from city lights to see it. It's what I call a stargazing deep track. In the early evening, it resides in the low southern sky. How that faint collection of stars is supposed to be the goddess of fertility, Demeter/Ceres, is a mystery to me.

OBSERVATION NOTES It only has one bright star, Spica, a blue giant star about 263 light-years, or about 1,546 trillion miles, away from Earth. It's 10 times as massive as our sun and three times the diameter of our home star. Spica is also a lot hotter than our sun, with a surface temperature well over 30,000 degrees Fahrenheit. On the northwestern corner of Virgo toward the constellation Leo is what's known as the Realm of the Galaxies or the Virgo Cluster. If you have a larger telescope, and you're really out in the country, you have a chance of seeing at least a few of the many galaxies, some over 60 million light-years from Earth! By the way, if you were to fly to the Virgo cluster in a jet airliner that flew at an average speed of 500 miles per hour, it would only take you about 80 trillion years to get there.

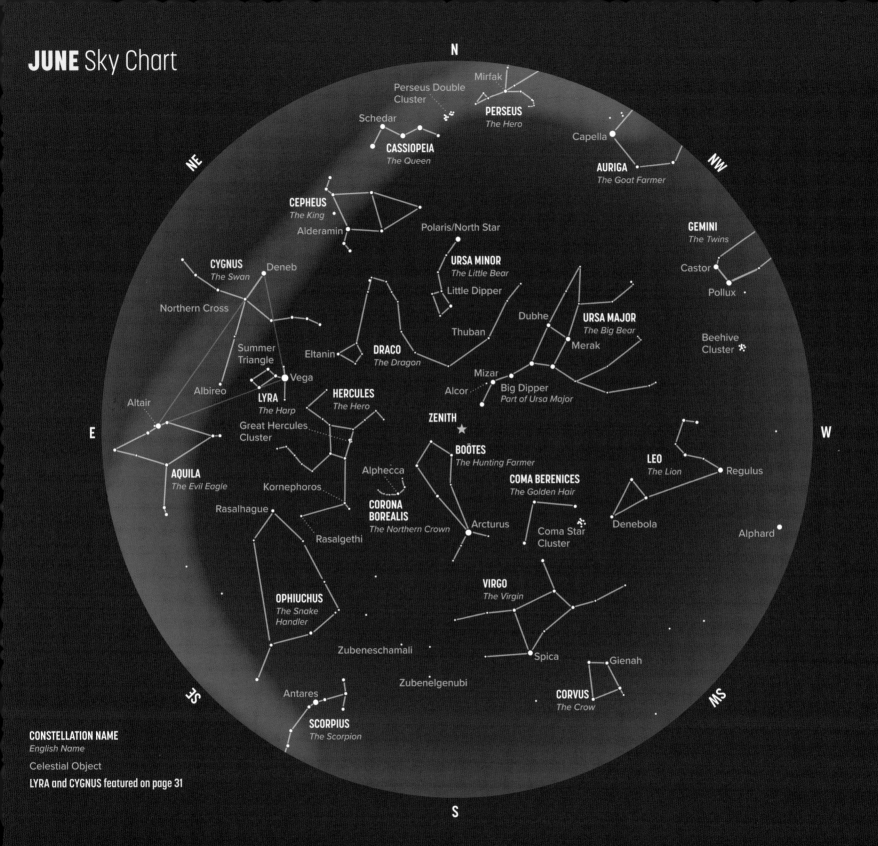

JUNE Sky Chart

N

Mirfak

Perseus Double Cluster

PERSEUS
The Hero

Schedar

Capella

CASSIOPEIA
The Queen

AURIGA
The Goat Farmer

NW

NE

CEPHEUS
The King

Alderamin

Polaris/North Star

GEMINI
The Twins

CYGNUS
The Swan

Deneb

URSA MINOR
The Little Bear

Little Dipper

Castor

Pollux

Northern Cross

Dubhe

URSA MAJOR
The Big Bear

Beehive
Cluster

Summer
Triangle

Eltanin

DRACO
The Dragon

Thuban

Merak

Albireo

Vega

Mizar

Big Dipper
Part of Ursa Major

Altair

LYRA
The Harp

HERCULES
The Hero

Alcor

E

Great Hercules
Cluster

ZENITH

W

AQUILA
The Evil Eagle

Alphecca

BOÖTES
The Hunting Farmer

LEO
The Lion

Regulus

Kornephoros

COMA BERENICES
The Golden Hair

Rasalhague

**CORONA
BOREALIS**
The Northern Crown

Arcturus

Coma Star
Cluster

Denebola

Rasalgethi

Alphard

OPHIUCHUS
*The Snake
Handler*

VIRGO
The Virgin

Zubeneschamali

Spica

Gienah

Zubenelgenubi

CORVUS
The Crow

SE

Antares

SCORPIUS
The Scorpion

SW

S

CONSTELLATION NAME
English Name

Celestial Object

LYRA and CYGNUS featured on page 31

Sky chart is representative of 10 p.m. in June, 2 a.m. in April, and 5 a.m. in February.

The stars of summer serenade you throughout the shortest nights of the year. To see them, consider pulling an all-nighter. It's worth being a little groggy.

The transition in the night sky is just about complete. The stars and constellations of winter are pretty much gone from our skies, all setting well before the sun. Of the winter constellations, only the bright stars Castor and Pollux are visible; part of the constellation Gemini, the Twins, these stars shine side by side in the low western sky just after evening twilight.

If you lie back on that reclining lawn chair and look straight overhead toward the zenith, you'll easily see the nearly upside down Big Dipper, and not far from the end of the Dipper's handle, you'll see a bright orange star. That's Arcturus, the brightest star in the constellation Boötes, the Hunting Farmer, which actually looks more like a giant nocturnal kite that has Arcturus at the tail.

Over in the eastern skies, the stars of summer are making their first appearance. Leading the way is Vega, the brightest star in the tiny constellation Lyra, the Harp. About all there is to Lyra is Vega and the stars that make up the parallelogram that represents the rest of the harp: Zeta 1 Lyrae, Delta 2 Lyrae, Sheliak, and Sulafat. They vary in distance from 154 light-years to over 900 light-years away. Even though they look like they're related, they just happen to fall in our line of sight to form that parallelogram.

Vega is a star almost 2.5 million miles in diameter, nearly three times the size of the sun. Vega is a much warmer star than our sun, however, with a surface temperature exceeding 16,000 degrees Fahrenheit. The reason Vega is so bright in our sky is twofold. First, it kicks out more than 50 times the light our sun does, and second it's a relatively close

star. Vega is 25 light-years away, which means the light you see from Vega this summer left that bright star a quarter of a century ago.

Scientifically, the most interesting thing about Vega is that, in the last part of the 20th century, both ground- and space-based telescopes discovered that Vega has a distinct dust disk surrounding it. That may be a sign of a developing solar system, and planets may already be circling Vega. In the past 30 years or so, astronomers have detected hundreds and hundreds of other planets around other stars; such planets are referred to as extra-solar planets, and thousands of candidate planets have yet to be confirmed. New discoveries are coming fast and furious!

If you have a small- to moderate-size telescope, the most interesting celestial treasure to find within the diminutive summer constellation Lyra is what astronomers call M57, the Ring Nebula. It looks like a tiny smoke ring, but with even a small telescope, you can see that M57 looks like a puffy star. You don't need skies as black as pitch to see it, but the darker the sky, the better. As you can see on the constellation diagram of Lyra (page 31), M57 is on the opposite side of the constellation from Vega, right between two of the stars that make up Lyra's parallelogram. See if you can spot it.

The Ring Nebula is categorized as a planetary nebula, but that moniker is a bit misleading. It really doesn't have anything to do with the planets orbiting in our solar system or any other planets. Planetary nebulae got that name because they were discovered in the 18th century with telescopes that were a lot smaller and not nearly as sophisticated as those in existence today. When viewed with that equipment, planetary nebulae looked similar to giant planets like Jupiter and Saturn. The Hubble Telescope and James Webb Telescope were still a few hundred years away from being available.

We now know that planetary nebulae are stars entering retirement and that will soon become white dwarfs. All stars, with the exception of behemoths, go through this process. Throughout most of their lives, stars produce light and energy through a process called nuclear fusion. Deep in a star's core, tremendous heat builds up because of gravitational compression, which simply means the star is being squeezed by its own gravity. The sun's core temperature is around 27 million degrees Fahrenheit.

Stars primarily consist of hydrogen atoms, and the heat causes hydrogen to fuse into helium atoms, which are heavier. The nuts and bolts of it are a little too involved to go into here, but it suffices to say that the process produces immense amounts of energy and light.

Eventually, a star runs out of hydrogen in its core and helium atoms begin to fuse into carbon and oxygen. Again, the details get a little complicated, but when a star can no longer fuse atoms in its core, it begins to collapse in on itself due to gravity. As it does, the star temporarily puffs out or burps out shells of its remaining gases as the core shrinks into a white dwarf. Those burping shells are what you see when you look at planetary nebulae. Our own sun will go through this process in roughly 6 billion years or so, and the resulting white dwarf won't be much larger than Earth. That's what's happening right now in the Ring Nebula, which is over 2,300 light-years away. The light that you see now from this ailing star has been traveling thousands of years just to meet your eyes!

A little to the lower left of Vega and the constellation Lyra is the moderately bright star Deneb, the brightest shiner in Cygnus, the Swan. Cygnus is otherwise known as the Northern Cross and rises sideways in the east. Deneb lies at the head of the cross and is over 1,500 light-years from Earth. Deneb is a moderately bright star in our sky, but looks are

deceiving. It's almost 60,000 times brighter than our sun and possibly over 90 million miles in diameter. Our own sun is less than a million miles in diameter. Deneb would be a whole lot brighter in our sky, except that it's so very far away. The starlight you see from Deneb this month left that star around 500 A.D.! Just the fact that you see Deneb as well as you can testifies to the size and brilliance of this star that marks the tail of Cygnus, the Swan.

Albireo, the star that represents the head of Cygnus, the Swan, sits at the base of the Northern Cross. By the way, take a small telescope and have a look at Albireo. You'll see that Albireo, which looks somewhat unimpressive with the naked eye, is actually a gorgeous, colorful double star. One of the stars is a distinct blue and the other a pale orange, with both stars shining from nearly 400 light-years away. You'll love it!

DUMBELL NEBULA

RING NEBULA

NORTH AMERICAN AND PELICAN NEBULA (right)

JUNE Featured Constellations

Numbers under star names represent light-years

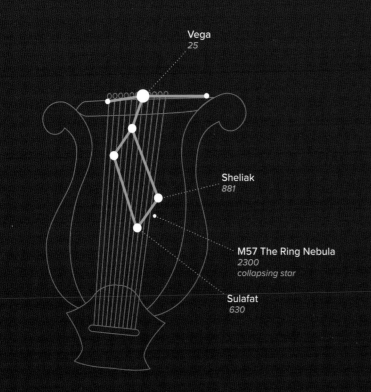

Vega
25

Sheliak
881

M57 The Ring Nebula
2300
collapsing star

Sulafat
630

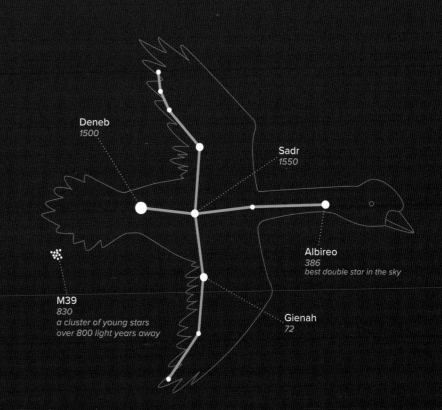

Deneb
1500

Sadr
1550

Albireo
386
best double star in the sky

M39
830
a cluster of young stars
over 800 light years away

Gienah
72

LYRA *The Harp*

BACKGROUND/MYTHOLOGY The Greeks considered Lyra to represent a lyre, an old-fashioned harp, in remembrance of the great musician Orpheus. Orpheus played his lyre so well that he could charm anyone. He used it to attract Eurydice, the love of his life. Sadly, Eurydice died at a young age, and Orpheus fell apart. Pluto, the god of the underworld, felt so sorry for Orpheus that he told Orpheus to lead Eurydice out of the realm of the dead on the condition that he was not to look back at her as they walked. Unfortunately, Orpheus grew afraid that Eurydice was not behind him and could not resist looking back. Eurydice was there but because of Orpheus's failure, she was banished forever. When Orpheus died, the rest of the gods of Mount Olympus placed his harp among the stars to make never-ending heavenly music.

OBSERVATION NOTES The constellation Lyra is faint but contains one of the brightest stars in the sky, Vega. As soon as it gets dark enough after sunset (around 10 p.m.), look for the brightest star you can see in the eastern sky. You can't miss Vega, a star over 25 light-years away. Then, look for a small parallelogram just to the lower right of Vega. That's all there is to Lyra.

CYGNUS *The Swan*

BACKGROUND/MYTHOLOGY Phaethon was the son of Apollo, the god of the sun. One early morning, before Apollo woke up, Phaethon broke into the barn where Apollo kept his flying horses and sun cart. He grabbed the reins and tried unsuccessfully to fly the sun across the sky. Apollo took a spare horse-drawn flying cart, grabbed the reins away from his son, and brought the sun cart under control. In the process, Phaethon fell to his death. The rest of the gods put a celestial swan in the sky as a memorial to Phaethon.

OBSERVATION NOTES In late spring, Cygnus, the Swan, is flying in the early evening eastern sky. Its brightest star is Deneb, one of the stars of the Summer Triangle, which resides in the eastern sky in the early evening. The other stars of the Summer Triangle are Vega and Altair, the brightest stars of their respective constellations, Lyra, the Harp, and Aquila, the Eagle. They are the brightest stars in that part of the sky. Looking east, Deneb marks the lower left-hand corner of the Summer Triangle. It's the dimmest of the triangle, but it's not small. On the contrary, it's huge. Deneb's diameter may exceed 90 million miles, and it may produce 50,000 times more light than our sun.

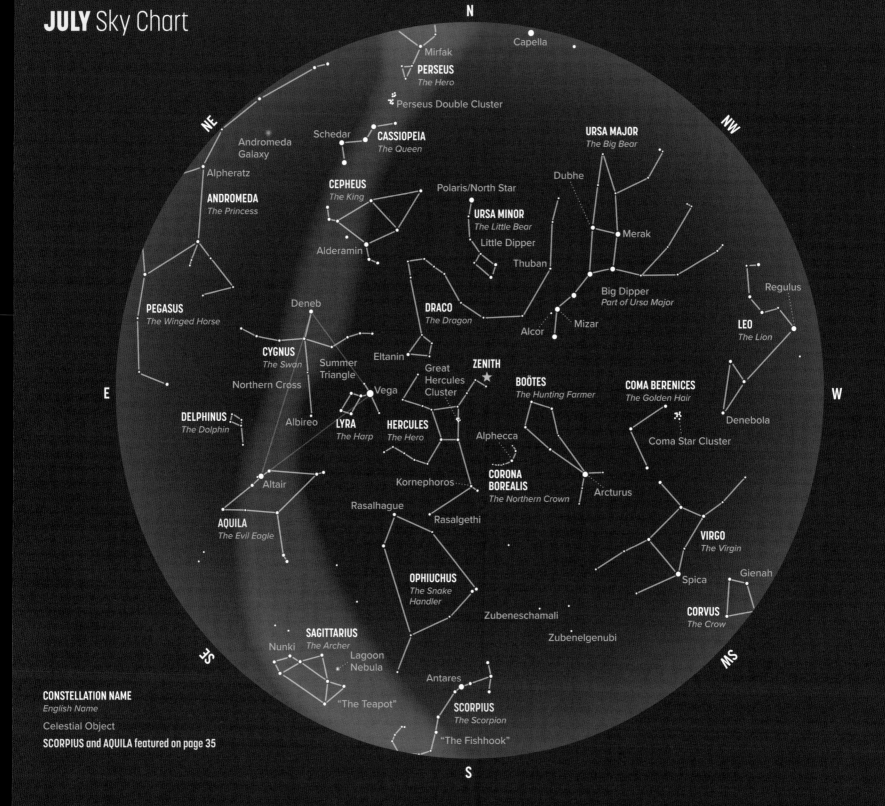

JULY Sky Chart

N

Capella

Mirfak

PERSEUS
The Hero

Perseus Double Cluster

NE

Andromeda
Galaxy

Schedar

CASSIOPEIA
The Queen

URSA MAJOR
The Big Bear

NW

Alpheratz

CEPHEUS
The King

Polaris/North Star

Dubhe

ANDROMEDA
The Princess

URSA MINOR
The Little Bear

Merak

Alderamin

Little Dipper

Thuban

PEGASUS
The Winged Horse

Deneb

DRACO
The Dragon

Big Dipper
Part of Ursa Major

Regulus

E

Eltanin

Mizar

LEO
The Lion

CYGNUS
The Swan

Summer
Triangle

Alcor

Northern Cross

Vega

Great
Hercules
Cluster

ZENITH ★

BOÖTES
The Hunting Farmer

COMA BERENICES
The Golden Hair

W

DELPHINUS
The Dolphin

Albireo

LYRA
The Harp

HERCULES
The Hero

Alphecca

Denebola

Coma Star Cluster

Altair

Kornephoros

**CORONA
BOREALIS**
The Northern Crown

Arcturus

AQUILA
The Evil Eagle

Rasalhague

Rasalgethi

VIRGO
The Virgin

OPHIUCHUS
*The Snake
Handler*

Spica

Gienah

Zubeneschamali

CORVUS
The Crow

SE

SAGITTARIUS
The Archer

Nunki

Zubenelgenubi

Lagoon
Nebula

Antares

SW

"The Teapot"

SCORPIUS
The Scorpion

CONSTELLATION NAME
English Name

"The Fishhook"

Celestial Object

SCORPIUS and AQUILA featured on page 35

S

Sky chart is representative of 10 p.m. in July, 2 a.m. in May, and 6 a.m. in March.

The stars and constellations of summer are now fully deployed in the late-night skies. The Summer Triangle, Scorpius, and Sagittarius are all waiting for you!

The stars and constellations of summer are out of their winter hibernation, and they're waiting for you to spend a warm, clear evening with them. July stargazing isn't quite as flashy as a Fourth of July fireworks show, but you can stargaze every clear night this month, and let's face it, the average municipal fireworks show only goes on for a half hour. Have the bug repellent ready, though, because in many parts of the country, mosquitoes love to get up close and personal with you. A lot of people get discouraged about summer stargazing because of these winged scourges of summer evenings, but in most cases, the bugs start backing off a couple of hours after dark as their bedtime feeding has pretty much ended.

In the northern July heavens, look for the Big Dipper in the northwestern sky and the fainter Little Dipper standing on its handle with Polaris, the North Star, at the end of the handle. Every single celestial object appears to revolve around Polaris every 24 hours. It's not just like clockwork. It is clockwork!

In the high southwestern sky, you'll see the brightest star in the sky tonight, Arcturus, which is also the brightest star in Boötes and the second-brightest star you can see throughout the course of the year. Boötes is supposed to be a farmer hunting the Big Bear. It's much easier to see it, though, as a giant kite with Arcturus at the tail.

In the eastern heavens, you'll see the prime stars of summer on the rise. The best way for finding your way around the summer stars is to locate the Summer Triangle, made up of three bright stars, the brightest in each of their respective constellations.

You can't miss them. They're the brightest stars in the east right now.

The highest and brightest star is Vega, the bright star in a small, faint constellation called Lyra, the Harp. On the lower left corner of the Summer Triangle is Deneb, the least-bright member of the Summer Triangle but still very easy to see. Deneb is a star more than 1,500 light-years away. It's also the brightest star at the tail of Cygnus, the Swan. Cygnus is also known as the Northern Cross because that's what it really looks like. Deneb is at the top of the Northern Cross, presently lying on its side as it rises in the east. Altair, located at the lower-right corner of the Summer Triangle, is the second-brightest star of the triangle and the brightest in Aquila, the Eagle. Altair is on the corner of a diamond that outlines the wingspan of the great constellation Aquila.

To avoid confusion among astronomers, back in 1922, the International Astronomical Union divided up the entire celestial dome into 88 constellations, which were mostly named after the tales from Greek and Roman mythology. Eight of these constellations are supposed to picture birds, and some are much better than others. Aquila, the Eagle, and Cygnus, the Swan, are two of the best of the bird constellations.

Altair itself is one of the more fascinating stars available to us Earthlings. It's one of the 15 brightest stars in the sky and relatively close, only 16 light-years (or about 97 trillion miles) away. Believe it or not, that's a lot closer than most stars you can see in the night sky. Because it's so close, astronomers know quite a bit about it. Altair is almost 1.5 million miles in diameter, twice as large as our sun, and it cranks out a lot more light than our home star, more than 10 times as much.

The most fascinating discovery made about Altair is that it has a bulging waistline. The Palomar Observatory in California discovered that Altair's diameter

is more than 20% larger across its equator than from pole to pole. Further observations revealed that Altair is rapidly spinning on its axis at the rate of one full rotation in less than 9 hours. By comparison, our sun takes more or less an entire month for one rotation. Altair, like all other stars, is basically a big ball of gas, so as it rapidly whirls, centrifugal force, the same force you feel on a fast merry-go-round, causes Altair to bulge out at its equator.

Scan your telescope all around Aquila, and you'll find some nice little star clusters of young stars, but one of the best sights through a small to moderate telescope is Messier Object 11, just off the tail of Aquila. M11, as it's often referred to, is technically in a small adjacent constellation called Scutum, the Shield. M11 is a beautiful open cluster of almost 3,000 stars and more than 6,000 light-years away. (That's 35,000 trillion miles!) This cluster is full of stars that are only 220 million years old, which is considered infancy in terms of stellar age. M11 has a nickname, the Wild Duck Cluster, because many people think it looks like a flock of flying ducks. Crank up your imagination to see that image!

In the low southern sky, there's a bright brick-red star called Antares that marks the heart of Scorpius, the Scorpion, one of the rare constellations that actually resembles what it's supposed to be. As evening begins it's about at its highest point in our southern skies, but that's not all that high. The Scorpion is a low-rider with its stinger barely above the horizon in most of the US.

Some people think Scorpius looks like a giant fishhook trolling in the low summer skies. When I was much, much younger, I remember my grandma pointing out "the big fishhook" from the dock of her cabin near Garrison, Minnesota. As it turns out, Grandma and the rest of us aren't all that crazy in seeing Scorpius as a giant fishhook. That's how many ancient Polynesian cultures also saw it, and

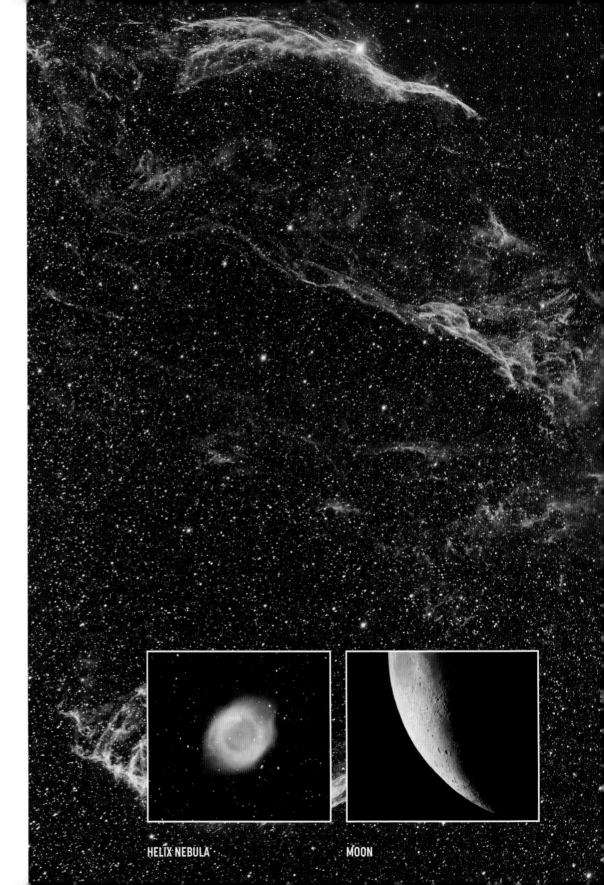

they saw the fishhook even better than we do from their much more southern locale. The farther south you are, the higher Scorpius is in the southern sky, with the exception of the part of the globe close to the South Pole.

Antares, a humongous star at the heart of the flying scorpion, is over 700 million miles in diameter and clearly demonstrates that stars aren't just little white lights. It has a distinctly ruddy-red hue to it. In fact, the name Antares is derived from Greek and roughly translates to "rival of Mars." Because of its ruddy hue, it can be easily confused with Mars. Many stars have a slight hue to them that can tell you a lot about them. Just like in a summer campfire, stars with reddish flames are relatively cooler than those with bluish flames. Our own sun is a yellowish-white star, and the temperature at its outer layer, called the photosphere, starts at a little over 10,000 degrees Fahrenheit. You won't need a jacket hanging out by Antares, but it is cooler at just under 6,000 degrees Fahrenheit. Antares is slowly dying as it's running out of nuclear fuel in its core. In the next billion years, or maybe even a little sooner, Antares will become so unstable that it'll blow itself to bits in a tremendous supernova explosion, and the residue from that great blast will spew out in all directions, becoming the building blocks of future stars and planets.

VEIL NEBULA (right)

HELIX NEBULA

MOON

JULY Featured Constellations

Numbers under star names represent light-years

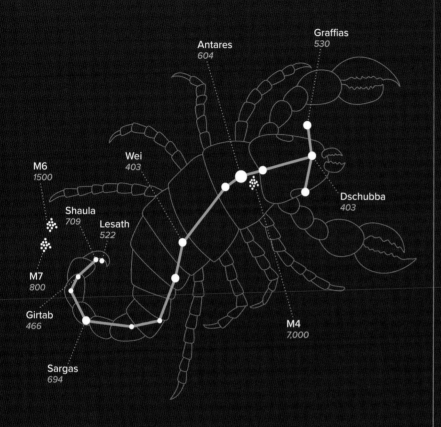

Antares
604

Graffias
530

Wei
403

M6
1500

Dschubba
403

Shaula
709

Lesath
522

M7
800

Girtab
466

M4
7,000

Sargas
694

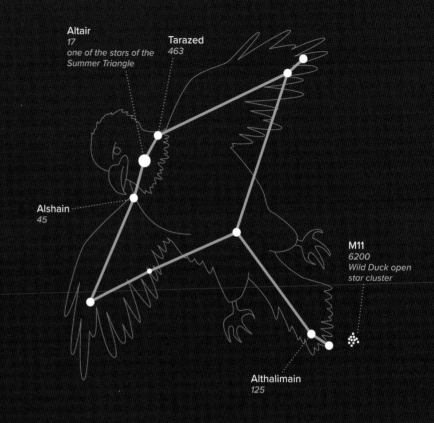

Altair
17
one of the stars of the Summer Triangle

Tarazed
463

Alshain
45

M11
6200
Wild Duck open star cluster

Althalimain
125

SCORPIUS *The Scorpion*

BACKGROUND/MYTHOLOGY Zeus's daughter, Artemis, fell in love with the great hunter Orion. Apollo, the brother of Artemis, greatly disapproved and sent a giant scorpion to kill Orion. After a long battle, the scorpion inflicted a fatal sting on Orion's neck before being crushed. The gods then placed both figures in the heavens.

OBSERVATION NOTES Scorpius is one of the truly great constellations of the summer. From its head to its stinger, Scorpius is one of those rare constellations that actually looks like what it's depicting. In July, Scorpius begins the evening in the low southern skies. In most of the US, you certainly won't need to crane your neck to see it. In fact, you really need to be observing in an area with an unobstructed view of the horizon to get a look at it, especially if you want to see its stinger. Even with such a vantage point, it's still a bit of a challenge because even with clear skies, visibility is naturally hampered closest to the horizon. Scorpius's brightest star, Antares, marks the heart of the scorpion. It has a dark red hue to it. It's what astronomers call a red supergiant star. And it's truly a behemoth! Our sun is about 864,000 miles in diameter, but Antares has a diameter of over 700 million miles! If Antares were placed in our solar system, the first four planets would be inside Antares!

AQUILA *The Evil Eagle*

BACKGROUND/MYTHOLOGY Zeus, the king of the Greek gods, was often a villain. Read up on Greek mythology, and you'll see what I mean. You never wanted him angry with you because he loved revenge. One of his tools of revenge was Aquila, his pet eagle. That bird was fearsome because it was trained to kill and kill slowly.

OBSERVATION NOTES The best way to find Aquila is to use the handy tool known as the Summer Triangle. This time of year, just look for the three brightest stars in the high eastern sky in the early evening. Each of these stars is the brightest in their own respective constellations. The highest and brightest star is Vega, the brightest star in Lyra, the Harp. On the lower left is Deneb, the brightest star in Cygnus, the Swan. The star on the lower right of the Summer Triangle is Altair, the brightest star in Aquila, the Eagle. As you can see on the diagram, Altair is at the upper-left corner of the large diamond that outlines the wingspan of the evil eagle. Altair is at the heart of the eagle. Across the diamond from Altair is a faint line of stars outlining Aquila's tail. The head of the eagle is to the upper left of Altair, but you'll have to totally rely on your imagination to see it as there are no stars outlining it.

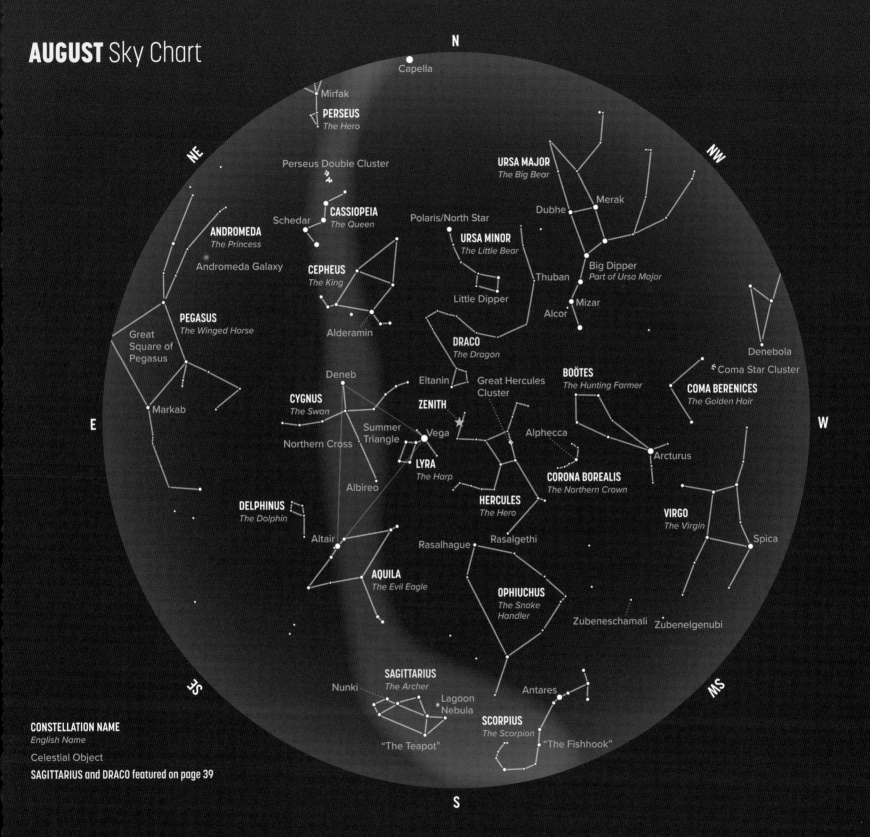

AUGUST Sky Chart

N

Capella

Mirfak

PERSEUS
The Hero

NE

Perseus Double Cluster

CASSIOPEIA
The Queen

Schedar

ANDROMEDA
The Princess

Andromeda Galaxy

CEPHEUS
The King

Polaris/North Star

URSA MINOR
The Little Bear

Little Dipper

Thuban

URSA MAJOR
The Big Bear

Dubhe

Merak

Big Dipper
Part of Ursa Major

Mizar

Alcor

NW

PEGASUS
The Winged Horse

Great
Square of
Pegasus

Alderamin

Markab

Deneb

DRACO
The Dragon

Eltanin

CYGNUS
The Swan

Northern Cross

Summer
Triangle

ZENITH

Vega

Great Hercules
Cluster

BOÖTES
The Hunting Farmer

COMA BERENICES
The Golden Hair

Denebola

♯ Coma Star Cluster

W

LYRA
The Harp

Albireo

Alphecca

CORONA BOREALIS
The Northern Crown

Arcturus

E

DELPHINUS
The Dolphin

Altair

HERCULES
The Hero

VIRGO
The Virgin

Spica

Rasalgethi

AQUILA
The Evil Eagle

Rasalhague

OPHIUCHUS
*The Snake
Handler*

Zubeneschamali

Zubenelgenubi

SE

SAGITTARIUS
The Archer

Nunki

Lagoon
Nebula

Antares

SCORPIUS
The Scorpion

"The Fishhook"

SW

"The Teapot"

CONSTELLATION NAME
English Name

Celestial Object

SAGITTARIUS and DRACO featured on page 39

S

Sky chart is representative of 10 p.m. in August, 2 a.m. in June, and 6 a.m. in April.

The splendid Milky Way band climbs across the top of the celestial dome to make for a magical night of stargazing you'll never forget!

August is awesome for stargazing. The summer constellations are in full bloom now. There's a lot of great stuff to gaze at and ponder. If you can, treat yourself by getting out to the countryside, or at least a little way away from the lights of the big city. Lie back on the ground with a blanket or take a load off in a reclining lawn chair. Forget about daytime stuff, like work, and just experience the universe.

High in the northwestern sky, you'll see the Big Dipper lazily hanging by its handle. The pot and handle of the Big Dipper actually represent the rear end and tail of the constellation Ursa Major, which is Latin for "Big Bear." See if you can spot a dimmer, skinny triangle of stars to the lower right of the pot. That's the Bear's head. Below the head and rear end, hunt for two curved lines of stars that make up his legs. You'll need a fairly dark site to see them. Not far from the Big Dipper, the Little Dipper (or Ursa Minor if you're into traditional Latin names) is upside down. Polaris, the North Star, is at the end of its handle.

In the eastern sky, the famous Summer Triangle is still holding court; it is made up of three bright stars: Vega, Deneb, and Altair. They are the brightest stars in that part of the sky, and each of them are the brightest stars in their individual constellations. Vega is the brightest star in Lyra (the Harp), Deneb is the brightest in Cygnus (the Swan, otherwise known as the Northern Cross), and Altair is the brightest star in Aquila (the Eagle).

In the low southern sky are two of my favorite constellations, and as far as I'm concerned, they're the signature constellations of summer. In the southwest is Scorpius, the Scorpion, with the bright brick-red star Antares at the heart of the scorpion. It's one of the rare constellations that resembles what it's supposed to. In the low southeastern sky is Sagittarius, which is supposed to be a half-man/half-horse shooting an arrow. Forget about that. Most people I know refer to it by its nickname, "the Teapot," because that's what it truly resembles.

Now, if you're viewing the celestial teapot in the countryside, you can clearly see a bright band of milky light stretching from the low southern sky all the way back over the celestial dome to the northern horizon. You're looking sideways into the disk of stars that make up most of the stars of our Milky Way galaxy. Our home galaxy is a spiral disk of stars more than 100,000 light-years in diameter but comparatively thin, only around a 1,000 light-years thick. When you see the Milky Way band you're actually looking into the plane of our galaxy. There are so many stars in that band, and they are so far away that all you see is their combined light all mashed together. The constellation Sagittarius is in the direction of the center of our galaxy, about 26,000 light-years away. The downtown section of our home galaxy would appear a lot brighter in our sky, but there's a lot of interstellar dust and gas in the way. Many astronomers believe that if it weren't for all that gas and dust, the sky around Sagittarius would be brighter than the full moon!

Nonetheless, the part of the Milky Way band around the Teapot is fairly bright anyway and loaded with a lot of fun stuff. Even with a small telescope or a pair of binoculars, you'll find many, many star clusters and nebulae. In fact, if it's dark enough where you're stargazing, with just the naked eye, you can see what looks like a puff of celestial steam above the spout of the teapot. That "puff" is the Lagoon Nebula, known astronomically as M8, and it is a bright emission nebula. It's one of the larger and brighter star factories you can see in the sky, and you don't need a very fancy telescope to get a really good look at it. It's basically a giant cloud of hydrogen, the raw material it takes to manufacture stars. It is roughly 100 light-years in diameter and around 5,000 light-years away. Within this cloud, many new stars are being born, some with solar systems and planets that could end up being like Earth.

My advice is to lie back on the ground or in that comfy lawn chair and get out your binoculars. Scan the entire band of the Milky Way and you'll run into some really pretty star clusters and other great stuff. At this time of year, galaxy gazing is at its best! Even though you can see a lot of the Milky Way, keep in mind that most of the stuff or matter that makes up our galaxy and other galaxies is invisible and simply dubbed "Dark Matter." Astronomers really don't know what it is, but stay tuned.

Another great thing about August stargazing is the chance to take in the Perseid Meteor Shower. If there's little or no moonlight to "whitewash" the sky, it's usually the best meteor shower of the year. The peak of the Perseids is right around August 12 and 13, but really, just about any night in the first half of August, you stand a chance to see at least some meteors or shooting stars. During the peak, there may be 50–100 per hour!

Meteors are often called shooting stars because they may look like stars, but they're actually space rocks breaking apart in the Earth's atmosphere. The broken pieces are usually between the size of a pebble or a fist. Most shooting stars and meteor showers are caused by debris left behind by comets that previously passed by the Earth. Comets are essentially dirty snowballs that partially melt when they get close to the sun. Debris from these partially melted comets is left in their wake and gravity between the particles keeps the debris trail intact. When Earth passes through this trail, you can see meteor showers. Meteor showers are best seen from midnight to just before morning twilight. This is because after midnight, you're now on the side of the Earth that's rotating right into the debris trail.

The debris trail that causes the Perseids is from Comet Swift-Tuttle, which comes by this part of our solar system about every 130 years and was last observed in 1992. There is some thought that Swift-Tuttle could possibly collide with the Earth in 2126, but that's been played down by a lot of astronomers.

In the meantime, tiny pieces of Comet Swift-Tuttle will slam into the atmosphere at speeds up to and over 40 miles per second and will easily be incinerated by air friction before they can get anywhere near you. In fact, most of the light you see from meteors as they streak across the sky is not caused by their fiery death but by ionization. These debris articles are zipping through our atmosphere so fast that the column of air they're going through becomes temporarily destabilized. Zillions of electrons are temporarily bounced away from the nuclei of zillions and zillions of atoms and that produces energy in the form of light. You can see the collective light of all this chaos.

The Perseids get their name because all the meteors seem to emanate from the general direction of the constellation Perseus, the Hero. In August, Perseus rises high in the northeastern sky in the early morning hours. Does that mean you just look toward the northeastern sky? Absolutely Not! If you do, you'll miss a lot of them. My advice for watching the Perseids or any other meteor shower is to lie back on the ground or a reclining lawn chair and look all around the sky. Watching a meteor shower with family and/or friends is a lot of fun because you have that many more eyes watching the big sky.

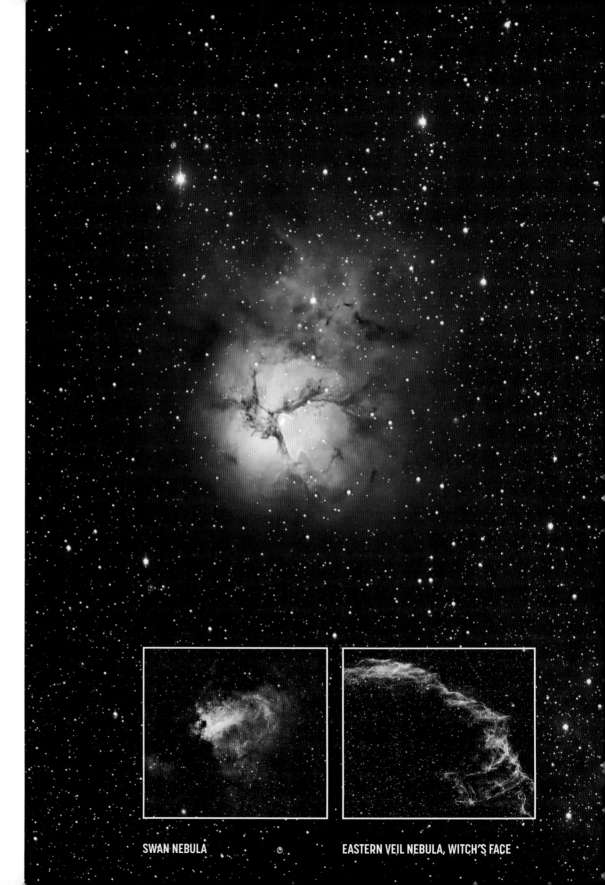

TRIFID AND LAGOON NEBULAE (right)

SWAN NEBULA

EASTERN VEIL NEBULA, WITCH'S FACE

AUGUST Featured Constellations

Numbers under star names represent light-years

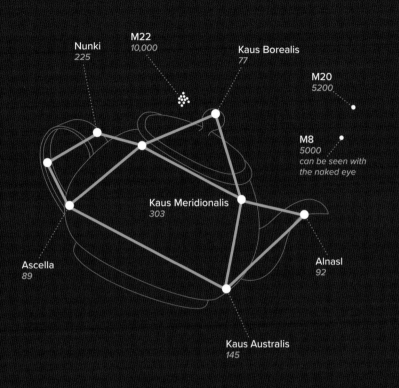

Nunki
225

M22
10,000

Kaus Borealis
77

M20
5200

M8
5000
*can be seen with
the naked eye*

Kaus Meridionalis
303

Ascella
89

Alnasl
92

Kaus Australis
145

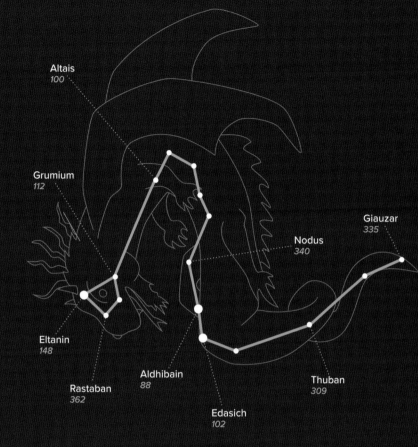

Altais
100

Grumium
112

Giauzar
335

Nodus
340

Eltanin
148

Aldhibain
88

Rastaban
362

Thuban
309

Edasich
102

SAGITTARIUS *The Archer*

BACKGROUND/MYTHOLOGY The southern end of the band of the Milky Way winds up in and around the constellation Sagittarius, the Archer. The ancient Greeks thought this constellation represented a centaur (a half-man/half-horse) shooting an arrow. What Sagittarius really looks like is a little teapot. In fact, the spout of the teapot points toward the direction of the center of the Milky Way. That part of the Milky Way would be much brighter in our skies if it weren't for a lot of interstellar clouds and dust blocking our view.

OBSERVATION NOTES One of the truly great, soulful things you can do for yourself is head out into the countryside and take in the starry skies at this time of year. When you do, you'll see a milky ribbon that cuts the sky in half. It's called the Milky Way Band and runs roughly from the northeast horizon to the southern horizon. All the stars you see in the night sky are part of the Milky Way, which may consist of 400 to 500 billion stars. Our galaxy is arranged in a spiral disk of stars 100,000 light-years in diameter but only 1,000 light-years thick. When you see the Milky Way Band, you're peering into the thinner plane of our galaxy. There are so many stars there that all you see is a combined milky glow.

DRACO *The Dragon*

BACKGROUND/MYTHOLOGY Draco isn't easy to see, but it's fun to find. It's certainly one of the larger constellations in the heavens, but its stars aren't all that bright. The best way to find Draco is to visualize it more as an uncoiled snake than a dragon. According to Greek mythology, Draco was the faithful "watchdog" of Hera, the queen of the gods. Tragically, Draco lost his life defending Hera and her precious golden apples. The intruder threw Draco's body into the sky so hard that his body contorted into the twisted constellation you see today.

OBSERVATION NOTES This time of year, the snake-like dragon is found in the high northern skies. Start out by looking in the high northern sky for the very bright star Vega, nearly overhead. Look a little to the right of Vega for a modestly bright trapezoid of four stars. That's Draco's head. From the head, hold your fist out at arm's length. At about two "fist widths" to the upper right, you'll find two faint stars fairly close to each other. These stars mark the end of the snake-dragon's neck. From those stars, Draco's body kinks downward and a bit to the left. Just above the Big Dipper's handle, the tail kinks off to the right and tapers off between the Big and Little Dippers. See if you can find the great unwound dragon!

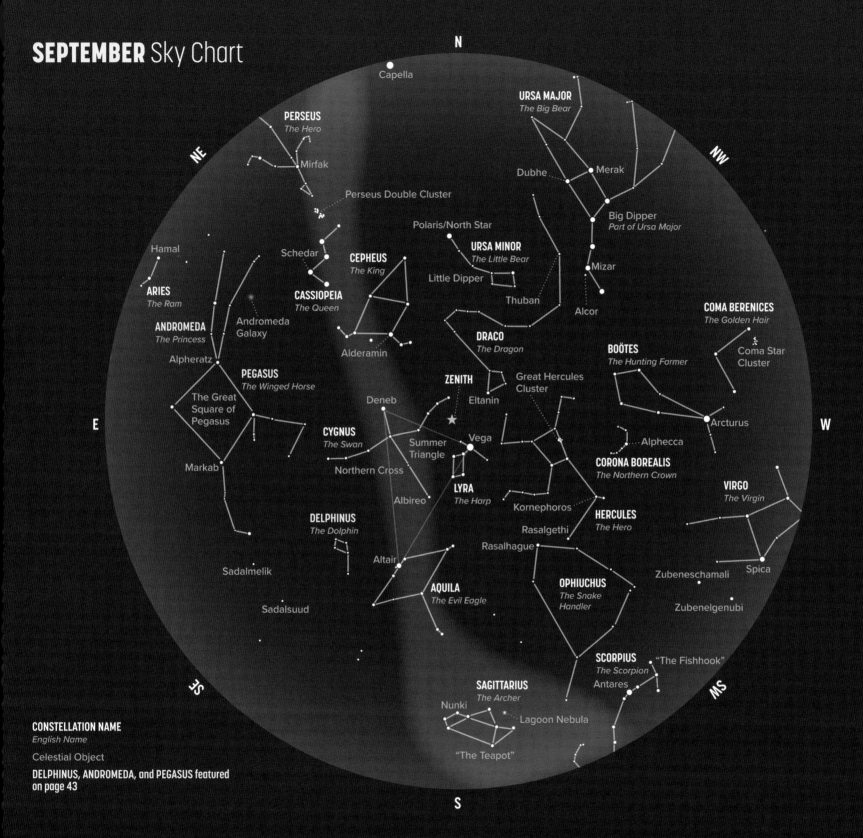

SEPTEMBER Sky Chart

N

Capella

PERSEUS
The Hero

Mirfak

NE

Perseus Double Cluster

Hamal

Schedar

CEPHEUS
The King

Polaris/North Star

URSA MINOR
The Little Bear

Little Dipper

Thuban

URSA MAJOR
The Big Bear

NW

Dubhe

Merak

Big Dipper
Part of Ursa Major

Mizar

Alcor

ARIES
The Ram

CASSIOPEIA
The Queen

Alderamin

ANDROMEDA
The Princess

Andromeda
Galaxy

Alpheratz

DRACO
The Dragon

COMA BERENICES
The Golden Hair

Coma Star
Cluster

PEGASUS
The Winged Horse

The Great
Square of
Pegasus

ZENITH

Eltanin

Great Hercules
Cluster

BOÖTES
The Hunting Farmer

E

Deneb

Vega

Arcturus

W

CYGNUS
The Swan

Summer
Triangle

Markab

Northern Cross

CORONA BOREALIS
The Northern Crown

Alphecca

VIRGO
The Virgin

Albireo

LYRA
The Harp

Kornephoros

HERCULES
The Hero

DELPHINUS
The Dolphin

Altair

Rasalgethi

Sadalmelik

Zubeneschamali

Spica

Rasalhague

AQUILA
The Evil Eagle

OPHIUCHUS
*The Snake
Handler*

Zubenelgenubi

Sadalsuud

SE

SCORPIUS
The Scorpion

"The Fishhook"

Antares

SW

SAGITTARIUS
The Archer

Nunki

Lagoon Nebula

CONSTELLATION NAME
English Name

Celestial Object

**DELPHINUS, ANDROMEDA, and PEGASUS featured
on page 43**

"The Teapot"

S

Sky chart is representative of 9 p.m. in September, 1 a.m. in July, and 5 a.m. in May.

The nights are getting longer, and you can start your great adventure under the stars before 10 p.m. Get away from the city lights, and you'll be hit broadside by the Milky Way Band that arches pretty much straight overhead. It's a ribbon of ghostly white light that makes up the thickest part of our home Milky Way Galaxy.

By September, the swan song of summer is loud and clear: schools are opening, leaves are starting to turn, and even a few Christmas decorations are going up in department stores. If you're like me, you hang on to summer as long as you can, and one way to do that is September stargazing. There are still plenty of summer constellations playing on stage in the celestial theater, and the nights are getting longer, so you don't have to stay up so late to take in the show!

One of my favorites is in the low southern to southwestern sky. I call it the Teapot. Now, for you purists, the Teapot is formally known as Sagittarius, a centaur shooting an arrow at its neighboring constellation to the west, Scorpius, the Scorpion. If you can manage to see Sagittarius as a half-man/half-horse with a bow and arrow, great! I'll stick with the Teapot.

The Teapot is located in the direction of the center of our Milky Way Galaxy, which is a little over 25,000 light-years away. If the sky is dark enough where you are, you'll see a milky white band of light that runs from the Teapot in the southwest sky all the way across to the northeast horizon. When you see that, you're looking at the combined light of billions of distant stars that make up the main plane of our galactic home.

Another signpost of summer, the Summer Triangle, is nearly overhead. Just look for the three brightest stars you can see around the zenith and that's it. All three stars are the brightest stars in each of their respective constellations. Vega is the brightest star in the constellation Lyra (the Harp), Altair is the brightest in Aquila (the Eagle), and Deneb is the brightest star in Cygnus (the Swan, which is also known as the Northern Cross.)

There's nothing really all that summery about the Big and Little Dippers since they're visible every night of the year, but summer is a great time to spot them. That's especially true for the Big Dipper, since it's proudly hanging by its handle high in the northwest. The fainter Little Dipper is standing on its handle to the right of the Big Dipper, with Polaris, the North Star, marking the end of the handle.

Even though there are summer constellations still in the sky, you can't deny that the seasons are changing. Autumn is coming (and eventually winter), but that's OK because as you forge deeper into September, the best autumn constellations are making a grand entrance in the eastern evening sky. In fact, September sky-watching is like a happy hour because it features a good two-for-one deal: Andromeda, the Princess, hitching a ride on the tail of Pegasus, the Winged Horse. Now, if you insist on the traditional depiction of this constellation, you are allegedly seeing Pegasus flying upside down with virtually no wings visible and tiny, faint little legs. I could take you outside and attempt to show it to you that way with my green laser pointer, but I guarantee you would be heavily underwhelmed.

Many stargazers (including me) prefer to see the same stars in a different way: we view it as a majestic flying horse with a huge wingspan that is upright and rescuing lovely Princess Andromeda from a ravenous sea monster. That's how I show it at all my star parties, and folks love it that way. In my depiction of Andromeda and Pegasus, Pegasus invades the traditional territory of Andromeda a bit, but in my opinion, it looks a lot more like what it's supposed to represent. After all, is there really such a thing as the "correct" interpretation of a constellation?

To see the flying horse of the heavens, first find what's known as the Square of Pegasus. Normally a square of four moderately bright stars, at this time of year, it actually looks more like a giant diamond in the eastern evening sky. (Some stargazers even call it the Autumn Diamond.) These stars outline the torso of the airborne horse. The star at the top right-hand corner of the diamond is Scheat, pronounced she—at. Scheat is the base of the horse's neck. Look above Scheat for two other stars that outline the rest of the neck. Then, look for another fairly faint star to the lower right of the neck that marks the flying horse's nose.

The horse has a front leg that is magically multi-jointed and extends upward in a curved line. To see it, start at Markab, on the right corner of the diamond. From there, look for a curved line of slightly fainter stars that extends up to the upper right of Markab. The modestly bright star at the hoof of Pegasus is called Enif.

I love the name of the star on the left corner of the diamond. It's called Alpheratz, pronounced Al-fee-rats. You can't help but see a curved line of three bright stars extending to the upper left of Alpheratz. It represents the mighty wings of Pegasus. If you look above that bright line of stars, you'll see another curved line of fainter stars. That outlines Andromeda, the Princess, who is hitching a ride on the back end of the horse. How the lovely princess found herself tied to the back of a flying horse is part of the great Greek mythological story involving Perseus, Cassiopeia, Pegasus, and the lovely Princess Andromeda. It's quite an epic!

Just above the curved line that outlines Andromeda, the Princess, is something that's truly out of this world. In fact, it's said to be the most distant object you can see with the unaided eye. What you're witnessing is the Andromeda Galaxy, the next-door galaxy to our home Milky Way. In areas with city lights, it's easy to miss; to have a good chance at spotting Andromeda, it's best to be in the countryside on a moonless night. Look for a faint misty patch of light just above the constellation Andromeda. If you're in an area with light pollution, binoculars or a small telescope will bring it into view.

The Andromeda Galaxy is more than two million light-years away. A light-year is defined as the distance that light travels in one year, so, by definition, the light you see emanating from the Andromeda Galaxy left our galactic next-door neighbor more than 2 million years ago.

But just how far away is it? Well, the speed of light is about 186,300 miles per second and one light-year equals about 5.8 trillion miles. If you do the math, that would put the Andromeda Galaxy at 2.2 million x 5.8 trillion miles away! Remember the Apollo spacecraft that took about three days to get to the moon and back in the late '60s and early '70s? Going at the same speed, it would take the Apollo capsule over 600 billion years to reach the Andromeda Galaxy! By the way, the Hubble Telescope has detected galaxies over 13 billion light-years away. It's no small universe, people!

FLAME NEBULA

ASDF

EAGLE NEBULA (right)

SEPTEMBER Featured Constellations

Numbers under star names represent light-years

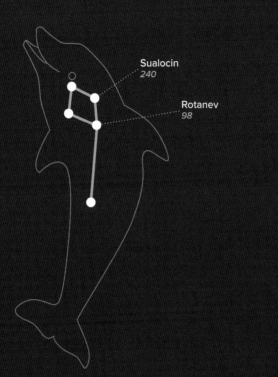

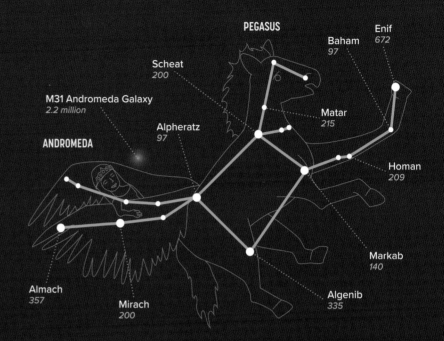

DELPHINUS *The Dolphin*

BACKGROUND/MYTHOLOGY There are many mythological stories involving this little sky dolphin, but the one I love best involves Arion, who was a poet and court musician. In this particular tale, he was sailing home from a concert when he was robbed by pirates and thrown overboard. He was permitted to play a song before being cast into the sea, and he played something so beautiful that the dolphin, Delphinus, saved him from drowning. Delphinus was placed in the stars as a reward for doing a good deed.

OBSERVATION NOTES Celestially speaking, Delphinus proves that great things come in small packages. Although it's tiny, it really resembles a dolphin. The best way to find Delphinus is by using the famous Summer Triangle that's still visible in September. It's very easy to see. Just look for the three brightest stars nearly overhead, and that's it. Each of these stars is the brightest in its respective constellation. The star on the lower southwest corner of the Summer Triangle is Altair, the brightest star in the constellation Aquila, the Eagle. From there, gaze just to the upper left of Altair and look for a small, dim, but distinct vertical diamond that outlines the head and torso of the diminutive dolphin. There's another equally dim star just below the diamond that marks the marine mammal's tail. In areas of light pollution, it can be a bit of a challenge to see, but in the countryside it's a cinch!

ANDROMEDA/PEGASUS *The Princess/The Winged Horse*

BACKGROUND/MYTHOLOGY Now, the great thing about constellations is that they're subject to interpretation. Traditionally, Pegasus is seen as a celestial horse flying upside down in the heavens with puny little wings. I don't see it like that, and I have lots of company. I see the giant horse proudly flying upright with a giant wingspan and towing the lovely Princess Andromeda.

OBSERVATION NOTES Pegasus, along with Andromeda, is one of the marquee constellations of the autumn and winter skies. You can't miss them in the eastern sky on September evenings. Just look for a giant tilted square or diamond of four stars; that's the Square of Pegasus, and it outlines the body or torso of the great horse. If it's dark enough where you're observing, you can spot the Andromeda Galaxy with the unaided eye, which is over 2 million light-years away. To do so, use the constellations Pegasus and Andromeda as your guide. The Andromeda Galaxy is just above the midpoint of the curved line that outlines Pegasus's wing and the fainter line that outlines Princess Andromeda. If it's dark enough, it'll look like a faint patch of light. With a telescope, you'll see a little bit more of our galactic next-door neighbor.

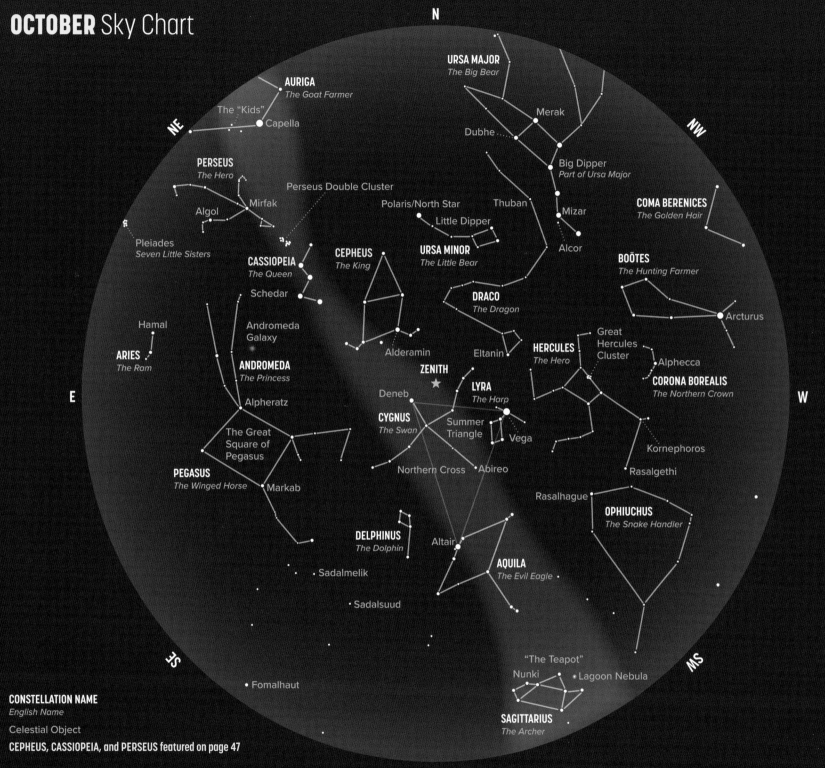

OCTOBER Sky Chart

N

URSA MAJOR
The Big Bear

Merak

Dubhe

Big Dipper
Part of Ursa Major

Thuban

Mizar

Alcor

NW

COMA BERENICES
The Golden Hair

AURIGA
The Goat Farmer

The "Kids"

Capella

NE

PERSEUS
The Hero

Algol

Mirfak

Perseus Double Cluster

Polaris/North Star

Little Dipper

URSA MINOR
The Little Bear

CEPHEUS
The King

BOÖTES
The Hunting Farmer

Pleiades
Seven Little Sisters

CASSIOPEIA
The Queen

Schedar

DRACO
The Dragon

Arcturus

Hamal

Andromeda
Galaxy

Alderamin

Eltanin

HERCULES
The Hero

Great
Hercules
Cluster

ARIES
The Ram

ANDROMEDA
The Princess

Alpheratz

ZENITH
★

Alphecca

CORONA BOREALIS
The Northern Crown

E

LYRA
The Harp

W

Deneb

The Great
Square
of Pegasus

CYGNUS
The Swan

Summer
Triangle

Vega

Kornephoros

PEGASUS
The Winged Horse

Markab

Northern Cross

Abireo

Rasalgethi

Rasalhague

OPHIUCHUS
The Snake Handler

DELPHINUS
The Dolphin

Altair

Sadalmelik

AQUILA
The Evil Eagle

Sadalsuud

SE

MS

"The Teapot"

Nunki

Lagoon Nebula

Fomalhaut

CONSTELLATION NAME
English Name

Celestial Object

CEPHEUS, CASSIOPEIA, and PERSEUS featured on page 47

SAGITTARIUS
The Archer

S

Sky chart is representative of 8 p.m. in October, 12 a.m. in August, and 4 a.m. in June.

It's a celestial Octoberfest in the night sky. While you have to put on a heavier coat to stargaze, October night skies have so much going for them. I call it a great stargazing harvest.

It's time to get out and enjoy the absolute beauty of the autumn night sky. We're entering one of the prime times for stargazing. The nights are longer, and with less moisture in the air, the skies are generally more transparent. Most of the mosquitoes have disappeared. Even if you're not a big-time stargazing fan, you owe yourself the treat of taking in the celestial happenings as you lie back on a reclining lawn chair. The dark skies of the countryside are best, but it's a great show, even right from an urban backyard.

The Big Dipper is upright and riding low in the northwestern sky. In fact, it's getting so low that it's hard to see if you have a high tree line. The Big Dipper is the most famous star pattern there is, but it's technically not a constellation. The Big Dipper is actually the rear end and the tail of the constellation Ursa Major, the Big Bear. It's also the brightest part of the Big Bear.

One of the pieces of star lore I love involves the Big Bear and the nearby constellation Boötes, the Hunting Farmer. Boötes actually looks much more like a kite than a farmer and has a bright, noticeably orange star (called Arcturus) at the tail of the kite. All summer long Boötes has been hunting down the Big Bear. He's finally put several arrows into the beast and that's why it's falling so low in our sky. In fact, the Big Bear is bleeding, and as the blood falls on the trees and bushes, it causes them to turn the red colors of fall.

In the western half of the heavens, what's left of the summer constellations are still hanging in there. The Summer Triangle stars Vega, Altair, and Deneb make it easy to find the constellations Lyra, the Harp; Aquila, the Eagle; and Cygnus, the Swan. Also, in that same part of the sky is one of my favorite little constellations, Delphinus, the Dolphin. Just look for a faint little diamond of four stars that outline the dolphin's body and another faint star next to the diamond that marks the mammal's tail.

Because of Earth's orbit around the sun, the nighttime side of the Earth faces different directions in space during the course of the year, determining which constellations you will see at a given time. Over the course of October, you'll notice that the stars in the western half of the sky start out lower and lower as evening begins. Eventually, by the end of October, some of the stars that were in the low western sky at the beginning of the month will already be below the western horizon before dark. As you lose constellations below the western horizon, you will see new ones rising higher and higher in the east. What goes around comes around, though, because as the Earth goes around the sun, the stars and constellations you lose in the west eventually show up again in the east some months later. I just love this natural cycle.

In the eastern half of the sky, Pegasus, the Winged Horse, starts out the evening much higher in the sky than it did in September. Pegasus is carrying Andromeda, the Princess, at the command of Perseus, the Hero. Without a tremendous amount of imagination, you can kind of see Perseus as a stickman whose head is made up of a faint triangle of stars. His heart is represented by Mirfak, the constellation's brightest star, which is just under 600 light-years away from Earth. Perseus's most interesting star is Algol; it marks the severed head of the monster Medusa. According to Greek mythology, Medusa was turned into a monster and then cursed so that even a passing glance at her turned you to stone. One of the deeds that made Perseus a Hero was cutting Medusa's head off.

Astronomically, Algol is an eclipsing binary star. It's actually two stars that revolve around each other in a very tight orbit, one a little brighter than the other. When the dimmer star passes in front of the brighter star every 2.8 days, Algol gets dimmer by 30% for a few hours. For that reason, Algol is also referred to as the Demon Star. Watch Medusa's blinking head as it brightens and fades, but be careful not to get turned into stone!

An absolute must-see in the constellation Perseus is the great Perseus Double Cluster, just above the hero's head and not all that far away from the neighboring constellation Cassiopeia. If you're lucky enough to be stargazing in the countryside, you can see it with the naked eye as a misty patch among the stars. If you have to put up with a little light pollution, it's still a wonderful target with a pair of binoculars or a small telescope. It's a pair of huge open star clusters made up of about 100 young stars each that were born anywhere from 10–20 million years ago. Believe it or not, that makes them stellar children. (By comparison, the sun is in middle age at nearly 6 billion years old.) At a distance of 7,000 light-years away, both clusters are really out there. One light-year equals about 6 trillion miles. Because they're 7,000 light-years away, the light that you can see from them left those clusters around 5,000 B.C.!

Later in the evening, there's another bright cluster of stars that rises right behind Perseus. It's found in the constellation Taurus, the Bull, and I know you've seen it. It's the Pleiades, otherwise known as the Seven Little Sisters. It kind of resembles a tiny Big Dipper. The Pleiades are so much brighter than the Perseus Double Cluster because they're a lot closer, at a distance of just over 400 light-years. If the Perseus Double Cluster were the same distance from Earth that the Pleiades are, you'd notice about a quarter of our night sky would be dominated by the clusters. It would really be something to see!

The October skies are so full of wonderful sights, and if you stay up late, the famed winter constellations rise from their summer slumber, ready for another season of dazzling.

PERSEUS DOUBLE CLUSTER　　　**TRIANGULUM GALAXY**

　　　ANDROMEDA GALAXY (right)

OCTOBER Featured Constellations

Numbers under star names represent light-years

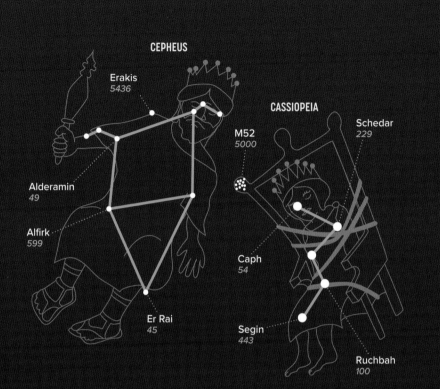

CEPHEUS

Erakis
5436

Alderamin
49

Alfirk
599

Er Rai
45

CASSIOPEIA

M52
5000

Schedar
229

Caph
54

Segin
443

Ruchbah
100

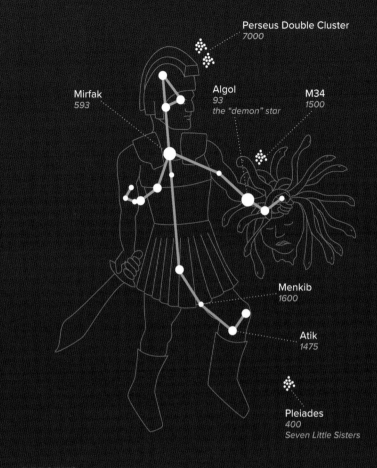

Perseus Double Cluster
7000

Mirfak
593

Algol
93
the "demon" star

M34
1500

Menkib
1600

Atik
1475

Pleiades
400
Seven Little Sisters

CEPHEUS/CASSIOPEIA *The King/The Queen*

BACKGROUND/MYTHOLOGY Cassiopeia was a powerful queen who once boasted she was more beautiful than even the nymphs of the sea. Poseidon sought to punish her because he had created the nymphs. He eventually tied Cassiopeia to her own throne and left her to hang in the sky for all eternity. Her husband, Cepheus, couldn't bear living without her, so Zeus, the king of the gods, placed him in the heavens next to his wife so they could be together forever.

OBSERVATION NOTES In October, the constellations Cassiopeia and Cepheus are side-by-side high in the northern sky in early evening. Cassiopeia, the Queen, is smaller than Cepheus but is much brighter. Just look for the sideways letter "W" that's tilting to the left, about to become an "M." That "W" outlines the throne and red carpet of Cassiopeia. The queen is tied up in her throne and is in eternal torment. King Cepheus begged Zeus, the king of gods, to be launched into the heavens so he could be with his beloved, and he's not as easy to spot. Rather than looking like a king, Cepheus really resembles a house with a steep roof and is nearly upside down. Look for Cepheus just to the left of Cassiopeia. To see it, look for the brightest star to the upper left of Polaris, the North Star; that's the apex of the roof of the house that represents Cepheus.

PERSEUS *The Hero*

BACKGROUND/MYTHOLOGY The constellation Perseus is one of the characters in the Greek mythological epic that involves King Cepheus, Queen Cassiopeia, Princess Andromeda, a winged horse, and a sea monster. Long story short: Perseus, a son of the god Zeus, was assigned to rid the countryside of the horrible monster Medusa. Instead of hair coming out of her head, Medusa had live snakes. She was cursed so that if you looked at her for even a millisecond, you turned to stone. Medusa had to be stopped because she was turning entire towns to stone. It was Perseus's mission to cut off Medusa's head and fly it back to Mount Olympus for disposal. To accomplish this, Perseus borrowed the winged shoes of Hermes, the messenger of the gods, in order to fly like a bird. He also borrowed the magic mirrored shield of the goddess Athena, which allowed him to lop off Medusa's head without gazing at it. As a reward, Perseus was placed in the stars, where he's still clutching Medusa's head.

OBSERVATION NOTES Be sure to check out the Double Cluster of Perseus, just above Perseus's head. With even a small telescope, you'll be blown away by the twin clusters of young stars that are over 7,000 light-years away.

NOVEMBER Sky Chart

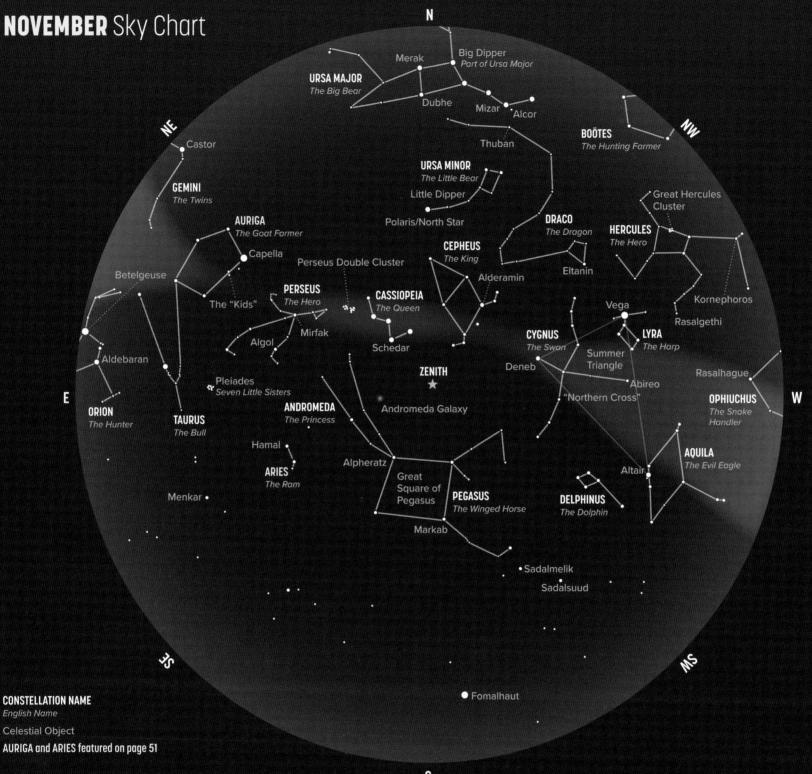

Sky chart is representative of 8 p.m. in November., 1 a.m. in September, and 3 a.m. in August.

There are summer constellations in the west, the wonderful winter constellations coming out of their summer slumber in the east, and, in between, the celestial wonders of autumn.

November skies have lots of thrills, but there are also a lot more chills. But if you're dressed for it, keep your feet warm, and have the right attitude, it's a perfect time to fall in love with the star-studded November skies. As a bonus this month, you get to set your clocks back an hour to standard time. With that, you get an extra hour of sleep, and you can start stargazing right after suppertime. You can go out as early as around 6 p.m.

In the high southern sky is the primo autumn constellation Pegasus, the Winged Horse, with Andromeda, the Princess, tagging along. Turn around and face north, and you'll see old friends like the Big Dipper, barely above the horizon, with the Little Dipper hanging by its handle higher in the northern sky. Cassiopeia, the Queen, the constellation that looks like a giant sideways "W," is in the high northeast sky. Cassiopeia's "W" outlines the queen's throne, and Cassiopeia is tied up in that throne. She angered Poseidon, god of the sea, by proclaiming that she was more beautiful than even the sea nymphs. So, Poseidon threw Cassiopeia into the sky and eternally bound her to a throne for all to see.

If moonlight doesn't light up the sky, one of the best celestial shows in November is the Leonid Meteor shower. It peaks in the pre-twilight hours on the morning of November 17. For a week or so prior to the 17th, you will see at least a few meteors, or shooting stars. Appropriately enough, the Leonids seem to emanate from the general direction of the springtime constellation Leo, the Lion, which is working the midnight shift in November.

By around 8 or 9 p.m., you will notice that there are a lot of bright stars on the rise in the eastern sky, and the later you stay up, the more of these wonderful winter constellations you can see. Orion, the Hunter, is the centerpiece. Orion is up by 10 p.m., but before then, you'll see a bright little cluster of stars in the low eastern sky, resembling a tiny dipper. It's not the Little Dipper, which is in the high northern sky.

What you're witnessing is the Pleiades Star Cluster, the best star cluster that you can see with the unaided eye. See how many stars you can spot. Can you see six? If you can, your eyes are about average. If you can see seven stars, you've really been eating your carrots! If you can see more than seven, you have super vision or you're just kidding yourself. A lot of you may know the Pleiades by its nickname, The Seven Little Sisters. Believe me, though, there are a lot more than seven stars there. With just an average pair of binoculars, you may see over a hundred stars!

Not far from the Pleiades is Auriga, one of the strangest constellations in the winter heavens. Auriga depicts a chariot driver with goats on his shoulder. To see it as advertised, you have to have a good imagination. Auriga is part of my favorite group of constellations, which I call "Orion and His Gang," and is perched above the mighty hunter's head in the low northeastern sky. Auriga resembles a giant lopsided pentagon with the bright star Capella at one of the corners. Capella is supposed to represent a mama goat perched on the retired chariot driver's shoulder. Capella is Latin and translates to English as "she goat."

Right next to Capella, see if you can spot the dim triangle of stars that represents Capella's baby goats. The baby goat star right next to Capella is over 2,000 light-years away. It looks dim and insignificant, but in reality, it's a big bright shiner a long way away.

Do you know the name of the brightest star you can see in the night sky most often? Yes, this is a trick question. It's not Sirius, the brightest star in the heavens. Actually, the brightest star you can see most often is Capella. Astronomically, Capella may seem like one bright star, but it's actually two huge stars separated by just over 60 million miles. That's less distance than between the Earth and the sun. Both of these stars are huge, each possibly over 10 million miles in diameter. Astronomers believe that these two behemoth stars orbit each other every hundred days or so.

Capella is one of the five brightest stars you can see throughout the course of the year, and it is the brightest star you can see most often in the northern hemisphere. That's because it's the closest bright star to Polaris, the North Star, which marks the position of the north celestial pole. In the northern hemisphere, stars and constellations that are close to Polaris, such as the Big and Little Dippers and the "W"-shaped Cassiopeia, are always above the horizon and travel in a tight circle around the North Star. Such stars are called circumpolar stars, and you can see them night after night. Capella, the goat star, is not quite near enough to Polaris to be considered a circumpolar star, but it's very close. Because of its northwardly position, Capella is present in our evening skies from late August until just about mid-June, and in most of the United States, it never goes a night without making a brief appearance.

In and around the pentagon that makes up Auriga are three small, somewhat faint, little, open star clusters, which are groups of young stars. Astronomically, these are referred to as M36, M37, and M38. "M" stands for Messier. Charles Messier was a French astronomer in the 1700s who was an avid comet hunter. Back then, telescopes weren't the quality instruments they are today. Throughout the skies, astronomers saw fixed fuzzy objects that were actually star clusters, nebulae, and even galaxies. Charles Messier didn't really care what they were, but he wanted to make a catalogue of them so he wouldn't mistake them for a comet, which also resemble little fuzzy patches. He knew these new shapes were comets if they changed their position from night to night against the background stars.

To this day, amateur and professional astronomers still refer to the Messier catalogue of 110 deep space objects. There are also more complete catalogues, like the New General Catalogue (NGC), the Index Catalogue (IC), and others.

ROSETTE NEBULA   CALIFORNIA NEBULA

PLEIADES STAR CLUSTER (right)

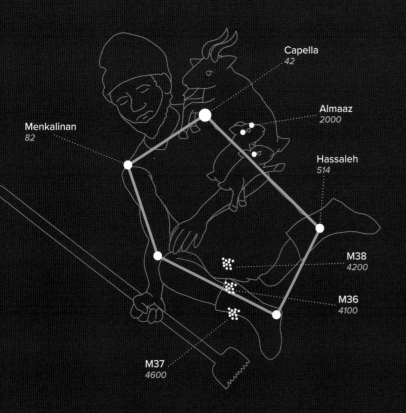

Capella
42

Almaaz
2000

Menkalinan
82

Hassaleh
514

M38
4200

M36
4100

M37
4600

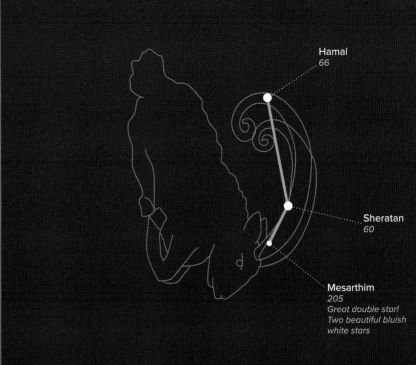

Hamal
66

Sheratan
60

Mesarthim
205
*Great double star!
Two beautiful bluish
white stars*

AURIGA *The Charioteer*

BACKGROUND/MYTHOLOGY One of the strangest constellations is Auriga, the chariot driver who is hauling goats around on his shoulder. According to some versions of Greek mythology, Auriga represents Myrtilus, the chariot driver for the mighty King Oenomaus. He betrayed the king for personal gain. When Myrtilus tried to collect his reward, he was double-crossed and killed. After his death, the gods placed his body in the sky where he became the constellation you can see during most of winter. It's not clear how Myrtilus wound up carrying goats, but my guess is that the imagination of shepherds tending their sheep at night had something to do with it.

OBSERVATION NOTES Auriga is part of my favorite group of constellations, which I call "Orion and His Gang," that is beginning to emerge above the eastern horizon. Auriga is one of the lead constellations in the low eastern November sky. Resembling a giant pentagon, Auriga has the bright star Capella at one of its corners. Capella marks the spot where a mama goat is parked on Auriga's shoulder. While Capella may look like one bright star, it's actually two stars that revolve around each other. The Capella system is 42 light-years away.

ARIES *The Ram*

BACKGROUND/MYTHOLOGY The Greek story of Aries, the Ram, is a sweet one. Aries sported a beautiful coat made of golden fleece and also had wings. One day, he was sent out to rescue two children from certain doom. He flew the children away from danger on his back, but only one survived the journey. In the end, Aries offered up his life as a sacrifice to save the remaining child. As a reward for his deeds, Zeus placed his image in the sky so that he would be remembered forever.

OBSERVATION NOTES Constellations come in all sizes, and Aries, the Ram, is definitely one of the smaller ones and honestly not all that dynamic. Aries is high in the eastern sky in the early evening this month. About all there is to it are two moderately bright stars, Hamal and Sheratan, and Mesarthim, a dimmer star that outlines the horn of the little ram. Aries used to be the backdrop constellation on the vernal (spring) equinox, which occurs on the first day of spring each year (around March 20). There are two equinoxes each year (one in spring, one in fall); at the equinox, the earth's axis is neither inclined toward the sun nor away from it. Because of Earth's slowly wobbling axis, Aries is no longer in the background at the vernal equinox; instead, Pisces (the Fishes) is.

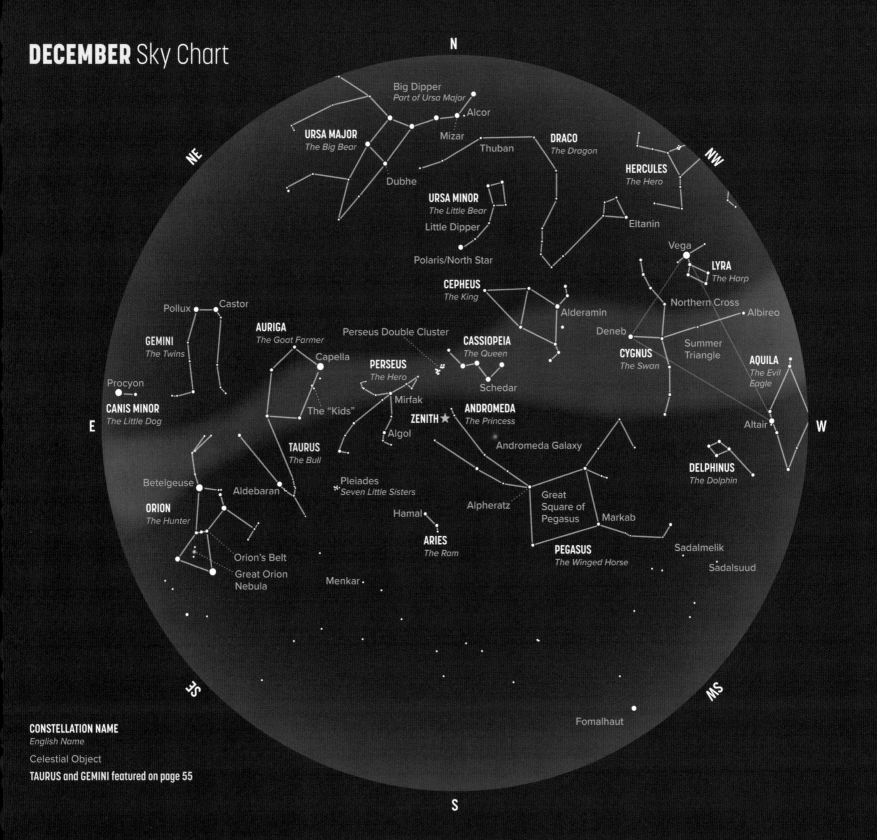

DECEMBER Sky Chart

N

Big Dipper
Part of Ursa Major

Alcor

Mizar

URSA MAJOR
The Big Bear

DRACO
The Dragon

Thuban

Dubhe

HERCULES
The Hero

NE

NW

URSA MINOR
The Little Bear

Eltanin

Little Dipper

Vega

LYRA
The Harp

Polaris/North Star

CEPHEUS
The King

Northern Cross

Albireo

GEMINI
The Twins

Pollux

Castor

AURIGA
The Goat Farmer

Perseus Double Cluster

CASSIOPEIA
The Queen

Alderamin

Deneb

CYGNUS
The Swan

Summer
Triangle

AQUILA
The Evil Eagle

Capella

Procyon

PERSEUS
The Hero

Schedar

CANIS MINOR
The Little Dog

The "Kids"

Mirfak

ANDROMEDA
The Princess

Altair

E

ZENITH ★

Andromeda Galaxy

W

TAURUS
The Bull

Algol

DELPHINUS
The Dolphin

Betelgeuse

Aldebaran

Pleiades
Seven Little Sisters

Great
Square of
Pegasus

Markab

ORION
The Hunter

Alpheratz

Hamal

Sadalmelik

Orion's Belt

ARIES
The Ram

PEGASUS
The Winged Horse

Sadalsuud

Great Orion
Nebula

Menkar

SE

MS

CONSTELLATION NAME
English Name

Fomalhaut

Celestial Object

TAURUS and GEMINI featured on page 55

S

Sky chart is representative of 8 p.m. in December, 1 a.m. in October, and 5 a.m. in August.

__The night skies have become as festive as the holiday season. Even before the holidays arrive, all of us are gifted with some of the best constellations of the year. Put on a heavy coat, grab a thick blanket and a comfy lawn chair, and check out the celestial holiday lights.__

'Tis the season for starwatching, but there's no sense sugarcoating it. Winter stargazing is not for hothouse flowers, but there is magic in the air during the holiday season. There's also magic in the heavens, as these long nights are blessed with some of the best constellations of the year. So, bundle up (maybe even wear a Santa hat!) and let the neighbors watching from a distance think you're off your rocker as you lounge on lawn chairs and gaze skyward. Then again, they're missing out on a great show, so why not invite them? Keep the hot coffee or hot apple cider (my favorite) handy and enjoy these special starlit nights.

With the longest nights of the year upon us, you can get out under the celestial delights by around 6 p.m., if not sooner. Believe it or not, despite winter's chill, there's still a touch of summer in the early evening western sky. Since June, we've enjoyed the constellation Cygnus, the Swan, which features Deneb, its brightest star, a blue supergiant star at least 1,500 light-years away. Several bright stars within Cygnus make up the famous Northern Cross, and during the holiday season, the cross is standing nearly upright above the northwestern horizon. By the way, I've seen the Southern Cross constellation (visible only in the tropics and the southern hemisphere), and I don't think it holds a cosmic candle to the grandeur of our Northern Cross.

The great winged horse Pegasus is riding high in the south-southwestern sky with Andromeda, the Princess, in tow. The Andromeda Galaxy, more than 2 million light-years away, is visible in the region; its light streams in from right around the zenith. With even a smaller telescope, it's an easy target this month. Cassiopeia, the Queen, is also in the same neighborhood and looks like a bright upside-down "W" or a right-side-up "M" in the high northern sky.

The later you stay up in the evening, the more you'll see of the best of the winter constellations that are rising in the east. By 8 to 9 p.m. you'll easily see Orion, the Hunter, that wonderful winter constellation. Its calling card is the row of three bright stars that make up Orion's Belt. A group of related bright constellations precedes Orion. There's Auriga (the Chariot Driver–turned–Goat Farmer); Taurus (the Bull), which features the wonderful Pleiades star cluster; and Gemini (the Twins), with the bright stars Castor and Pollux marking the foreheads of the twins. Gemini is such a famous constellation that it's mentioned in the Bible. In Chapter 28 of the Acts of the Apostles, it's written that the Apostle St. Paul climbed aboard a pagan ship bearing the figures of Castor and Pollux. Other constellations (such as Orion) are also referred to in the Good Book.

Located 50 light-years away, Castor is one of the most astronomically interesting stars on the celestial stage. To the naked eye, it looks like a single star, but with modern telescopes, astronomers have determined that Castor is actually a collection of five to six stars all orbiting around each other in a complex pattern. If you lived on a planet in that system, you would have six sunrises and six sunsets. Chances are that planets would have a hard time surviving in the gravitational chaos. There's a good chance that any planets in that tangled gravitational mess would be shot out into interstellar space to live the life of planetary orphans.

Pollux is a puffed-up version of our sun. Two times as massive as our sun, but at least eight times the sun's diameter, it is almost 7 million miles in diameter. If you placed Pollux in our solar system, you would need one thick pair of dark sunglasses: It produces 30 times more light than our sun! Thankfully, Pollux is quite a bit farther away than the sun, at a distance of over 35 light-years.

In Gemini, the best target for a small to moderate telescope is Messier Object 35, or M35 for short. It's a beautiful open cluster of young stars that occupies about the same area as the full moon in our sky. It's easy to find, right next to Castor's foot. It's over 2,500 light-years away and only 100 million years old, making the stars in this cluster stellar toddlers. M35 is a must-see with your scope.

By the way, this time of year it's especially important to let your binoculars or telescope (and its eyepieces) acclimate to the colder temperatures, or you might have less than desirable views. To do so, let them sit outside for at least a half hour before you start your observing session. Even if you do all this, you still may find that your views are unusually fuzzy, especially familiar targets you view all the time. When that happens, it's not your telescope's fault, but rather Earth's atmosphere's. Even if the skies seem perfectly clear, high winds (such as the strong jet stream) cause what's been dubbed "bad seeing." If the stars are really twinkling, that's a good sign of a stirred-up atmosphere. If that's the case, your best bet is to call it a night and try again when skies are clear, and the atmosphere is hopefully calmer and more transparent. As with any hobby or passion, you have to learn patience.

No matter if you're out with your telescope or binoculars or you're just sitting back on a lawn chair taking in the show, stargazing is wonderful for the soul. It always gives me a sense of awe, not only with the immensity of the universe but also how we're moving through it. First off, our world is rotating on its axis at over 1,000 miles per hour along the equator. At the same time, Earth is whipping along in orbit around the sun at over 66,000 miles per hour, or about 18 miles per second. But you also have to get your brain around the reality that the entire solar system is zipping along at nearly 600,000 miles per hour in a huge orbit around the center of the Milky Way. Now, if that's not enough motion for you, as you take in the show, remember that astronomers now believe that our entire Milky Way galaxy is cruising through this part of the universe at over 1.3 million miles per hour! Have you ever thought to yourself, "I'm going nowhere"? Well, nothing could be further from the truth.

CHRISTMAS TREE AND CONE NEBULA (right)

M35 CLUSTER

PELICAN NEBULA

DECEMBER Featured Constellations

Numbers under star names represent light-years

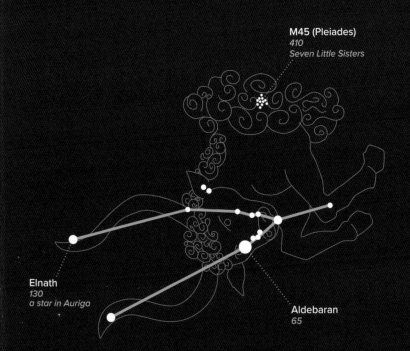

M45 (Pleiades)
410
Seven Little Sisters

Elnath
130
a star in Auriga

Aldebaran
65

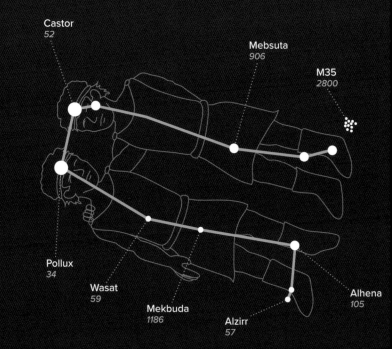

Castor
52

Mebsuta
906

M35
2800

Pollux
34

Wasat
59

Mekbuda
1186

Alzirr
57

Alhena
105

TAURUS *The Bull*

BACKGROUND/MYTHOLOGY Zeus, the king of the gods, had his eyes on the beautiful Princess Europa, who loved to raise prized bulls. Knowing this, Zeus disguised himself as a bull and hid among her livestock. When she tried to take him out for a ride, Zeus went wild. When he revealed his true identity, Europa fell in love with Zeus, at least for a time.

OBSERVATION NOTES Now, Taurus doesn't have a red nose like Rudolph, and as far as I know he's never guided Santa's sleigh, but Taurus does have a bright red eye, and it can help guide you around the winter sky. Taurus is easy to find. Just gaze toward the east. Right away you'll see there are a lot of bright stars in the eastern sky. You're seeing the rise of the winter constellations, the brightest of the year. You can't miss the bright and dazzling Pleiades star cluster at the heart of Taurus. It's a group of young stars and is absolutely dazzling through binoculars or a small telescope. Just below the Pleiades is a small arrow of stars that outlines Taurus's snout. Aldebaran, the brightest star on the lower side of the arrow, marks the red eye of the angry bull. If you extend both sides of the arrow to the left, you'll see the stars that mark the tips of the bull's long horns.

GEMINI *The Twins*

BACKGROUND/MYTHOLOGY In Greek mythology, Gemini is a touching tale of brothers—Castor and Pollux—who are loyal to each other in this world and in the next. Fittingly, Gemini's brightest stars are called Castor and Pollux, and they mark the heads of the twins. In China, Castor and Pollux are seen as a reminder of Yin and Yang, the two opposite aspects of nature. Even in areas with moderate light pollution, you can see two semicrooked but virtually parallel lines of stars leading down to the lower right of Castor and Pollux. If you're stargazing in an area with really dark skies, more stars are visible, and it looks like the two brothers are linking arms.

OBSERVATION NOTES Referring to Gemini as "the Twins" is a little redundant because gemini is the Latin word for "twins." Part of "Orion and His Gang," the Twins are perched in the low eastern sky to the left of Orion, and it's easy to see the celestial stickmen. Astronomically, Castor and Pollux are anything but twins. Pollux is an orange-red giant star over eight times the diameter of our sun and nearly twice its mass. Castor is actually at least a quadruple star system with possibly five or six stars, all in chaotic orbits around each other. When it comes to the night sky, looks can be deceiving.

ABOUT THE AUTHOR

Mike Lynch is a native Minnesotan who grew up in Richfield, Minnesota, in the 1960s. He attended Saint Peter's Catholic School and Holy Angels High School. After 2 years at the University of Minnesota in the Twin Cities, he transferred to the University of Wisconsin–Madison and earned his B.S. degree in Meteorology in 1979. Shortly afterward, he was hired as a broadcast meteorologist at WCCO Radio. He retired in 2020 after a 40-year career. Mike has covered all kinds of weather, from deadly tornadoes to record cold snaps. In fact, on February 2, 1996, he broadcasted from Tower, Minnesota, when the temperature dropped to 60 degrees below zero, an all-time record low for the state of Minnesota.

Since he was a teenager, Mike's other passion has been astronomy. He built his first telescope when he was 15 years old. For over 50 years, Mike has been teaching classes and hosting star parties with community education groups, nature centers, and other entities throughout Minnesota and Western Wisconsin. His goal is to help people make the stars their old friends.

From 2004 to 2007, Mike wrote astronomy/stargazing books for Voyageur Press. He wrote state-specific books for Minnesota, Wisconsin, Iowa, the Dakotas, Michigan, New York, Ohio, Pennsylvania, New England, New Jersey, the Carolinas, Florida, Georgia, Texas, Wyoming, Washington, Oregon, California, New Mexico, Arizona, Montana, and Southern Canada.

In 2007, he also wrote *Mike Lynch's Minnesota WeatherWatch*, which was a finalist in the Minnesota Book Awards.

On top of that, Mike also writes a weekly Starwatch column for the *St. Paul Pioneer Press*. His column is syndicated in several other newspapers across the United States.

Mike lives in Eagan, Minnesota, with his wife Kathy. He has two children, Angie and Shaun. Mike also enjoys astrophotography and has shared many of his better photos in this book.

ABOUT ADVENTUREKEEN

We are an independent nature and outdoor activity publisher. Our founding dates back more than 40 years, guided then and now by our love of being in the woods and on the water, by our passion for reading and books, and by the sense of wonder and discovery made possible by spending time recreating outdoors in beautiful places. It is our mission to share that wonder and fun with our readers, especially with those who haven't yet experienced all the physical and mental health benefits that nature and outdoor activity can bring. #bewellbeoutdoors